PRAISE FOR

SMALL TOWN, BIG SECRETS
INSIDE THE
BOCA RATON ARMY AIRFIELD
DURING WORLD WAR II

The Boca Raton Army Air Field was unquestionably one of the most significant places and times in Boca Raton history and had a major impact on the development of our community.

Susan Gillis
Archivist, Boca Raton Historical Society

Bravo to Sally Ling for telling the stories, some which have never previously been told, about Boca Raton Army Airfield. She has captured the human drama as well as the important contribution the base played during World War II. The early bombing campaign over Europe was less than effective due to overcast winter conditions encountered on most missions. By 1944, the advancement of radar bombing that was perfected on the flight line and classrooms of BRAAF turned the tide in the skies over Germany and Japan. The enhanced radar bombing accuracy probably shortened the war and saved the lives of countless aircrews.

Walter E. Houghton
Aviation Historian and Pilot
Deputy Director, Ft. Lauderdale-Hollywood International Airport

SMALL TOWN
BIG SECRETS

INSIDE THE
BOCA RATON ARMY AIRFIELD
DURING WORLD WAR II

SALLY J. LING

the
History
CHARLESTON PRESS LONDON

Published by The History Press
18 Percy Street
Charleston, SC 29403
866.223.5778
www.historypress.net

Cover image: Captain Harry Fromme, squadron commander (on left, first one standing), poses with the initial pilots assigned to the Boca Raton Army Air Field. In the background is a B-24. It is believed the two white lines on the nose were made by the photo lab, marking out the number of the aircraft for security reasons. *Photo courtesy of Lt. Col. Manuel Chavez USAF (Ret.).*

Front cover inset: The provost marshal's office at the main gate. *Photo courtesy of the Boca Raton Historical Society.*

Back cover inset: Lieutenant Jacob Beser poses in front of the *Enola Gay* at Roswell, New Mexico. The symbols on the plane are "Fat Men" and indicate missions the plane flew. *Photo courtesy of Leon Smith.*

First published 2005

Manufactured in the United Kingdom

ISBN 1.59629.006.4

Library of Congress Cataloging-in-Publication Data

Ling, Sally J.
  Small town, big secrets : inside the Boca Raton Army Air Field during
World War II / Sally J. Ling.
    p. cm.
  Includes bibliographical references and index.
  ISBN 0-596-29006-4 (alk. paper)
  1. Boca Raton Army Air Field (Fla.)--History. 2. World War,
1939-1945--Aerial operations, American. 3. Radar Military
applications--United States. 4. Radar--Research--History. I. Title.
  UG634.5.B6L56 2005
  940.54'43759'32--dc22
                    2005014509

Notice: The information in this book is true and complete to the best of our knowledge. It is offered without guarantee on the part of the author or The History Press. The author and The History Press disclaim all liability in connection with the use of this book.

Dedicated to the men and women who served on the Boca Raton Army Air Field.

CONTENTS

PREFACE

I moved to Boca Raton, Florida, in 1976. At the time, Interstate 95 stopped at a two-lane, east-west thoroughfare called Glades Road. Boca West was my home for the first year as I worked for the Arvida Corporation as director of club and resort activities. When I moved to Boca Linda West a year later, I would drive after work with my little dog, Pippin, to the campus of Florida Atlantic University. There, we would run on the tarmac of what I was told was once a World War II army airfield. While my father was a B-17 pilot in England during World War II and I grew up a military brat with some knowledge of military history, I had never heard of the Boca Raton Army Air Field. Little did I know thirty years later I would write a book about it.

It's been a glorious journey, one filled with awe and surprise. Awe in that I didn't realize Boca Raton had such a rich heritage—one steeped in mystery and wonderful memories. Surprise because I couldn't understand why someone hadn't told this amazing story before. But as I started my research, that reason became abundantly clear.

Information about the Boca Raton Air Field did not come neatly packaged like a gift to be opened. Instead, it came disguised as an onion—a big Vidalia onion. As I peeled away each sweet layer, more and more layers were revealed. Because there weren't any books written on the subject, it was necessary to review the few relevant articles in newspapers and periodicals, comb through reams of historical military documents, converse with military and civilian experts and historians, and locate dozens of veterans who once served at the field.

Many leads pointed to former military bases and archives where additional information could be gleaned, yet much documentation had been destroyed and most of the historians I spoke with had not heard of the Boca Raton Army Air Field. But it's no wonder. In those days, when secrets really were secrets, the Boca Raton Army Air Field was top secret. No one talked about it, therefore no one knew about it. The old military saying—the need to know—was very much alive and well; our nation's future depended upon it.

While this book has peeled off many of the layers surrounding the Boca Raton Army Air Field, I firmly believe many more are yet to be revealed. With a few more years, I might be able to get to the onion's core.

Without the Internet, this project could not have been pulled together so quickly nor told in such depth. The veterans would not have been able to share their wonderful stories; photographs could not have been acquired; documents and information not obtained and verified; nor historians and experts located. So to me, the production and timing of this book is nothing short of divine providence.

The story of the Boca Raton Army Air Field is organized by year but descriptions by veterans don't fall neatly into the year in which they occurred. While I tried to keep events as true as possible to when they transpired, many memories have been grouped by topic to support a cohesive story line. And, I did my best to make the story as accurate as possible, given the available information and personal remembrances.

The thunder of B-17 bombers is now but a whisper and the need for secrecy that surrounded the Boca Raton Army Air Field faded long ago, but in this book, I hope to preserve those memories—memories from an important time not only in world history but in the history of Florida, the town of Boca Raton and Florida Atlantic University.

Thank you for allowing me to share this extraordinary journey with you. I hope you enjoy the ride.

Sally J. Ling

ACKNOWLEDGEMENTS

No book is written without the assistance of others, whether it's research, technical support or simply encouragement. Allow me to take this opportunity to publicly acknowledge and thank those whose assistance enabled the story of the Boca Raton Army Air Field to be told.

First, I thank my husband, Chuck, who supported me unconditionally while I pursued this project with abandon.

Next, several people were especially instrumental in the production of this book. To them I owe a great debt of gratitude. They are: Sue Gillis, archivist for the Boca Raton Historical Society, for her outstanding knowledge, skills and technical support; the Boca Raton Historical Society for use of their documents and photographs; Amy An, Florida Atlantic University student and past president of Phi Alpha Theta, for her enthusiasm, incredible assistance in research, editing and willingness to go beyond the call to make this project happen; and Pat Keeley, my sister, who lent her unwavering support, a listening ear, superb editing skills and title suggestion. For their technical assistance, I thank Colonel Kenneth W. Davey USAF (Ret.), my father and former WWII pilot; Lieutenant Colonel Manual Chavez USAF (Ret.), former Boca Raton Army Air Field veteran; and Walter Houghton, acting deputy director, Fort Lauderdale–Hollywood International Airport.

Thanks also to Dick Randall for developing the maps; M.J. Beasley, collection development librarian at the Boca Raton Public Library, who helped with research materials; Boca Pioneers, many who shared their memories of Boca Raton; and Susan Bryant and Marci Shatzman, *Sun-Sentinel* editors who worked with me on the initial assignment that became the catalyst for this project.

Finally, I thank the veterans and civilians who shared their wonderful memories and photos of the Boca Raton Army Air Field. This is their story.

INTRODUCTION

From its inception, the Boca Raton Army Air Field was steeped in secrecy. Still today, some secrets exist. But it's precisely these secrets that give this story its flavor, and while many of these secrets will be revealed, some will remain hidden. Maybe forever.

The story of the Boca Raton Army Air Field does not begin in Boca Raton, Florida, with the October 15, 1942 official opening of the secret military installation, as some might believe. Nor does it begin in Washington, D.C., years earlier with discussions by top military brass regarding the construction of new bases to support the war effort. Instead, it begins in England with the invention of a device called RADAR (Radio Detection and Ranging)—a device that changed the world.[1]

While the German physicist Henrich Hertz began experimenting with radio waves in 1887, it wasn't until 1935 that British physicist Sir Robert Watson-Watt produced the first practical radar system. By 1940, Germany, Great Britain, France and the United States all used radar to navigate ships and planes as well as to detect enemy craft. But it was the invention of the next generation of radar that instilled the greatest hopes of positioning Britain ahead of Germany and allowed the Allies to win the war.

Radar changed the world, and it changed Boca Raton forever. From a sleepy town of around 700 in 1940, to the thriving Boca Raton Army Air Field that housed more than 100,000 troops from 1942 to 1947, to the 195,000 residents who inhabit Greater Boca Raton today, it was radar that played the central role in this makeover.

If we are to truly understand the effect radar had on the transformation of Boca Raton, it is important we take a step back to 1940 where the stage was being set by British and American scientists, military experts and politicians

---

1. (Since its invention, the word RADAR has become more common and is now lowercase, even though it is an acronym.)

1940

Secret Beginnings

Edward "Taffy" Bowen threaded his way through the crowded Euston Station in London, England, on August 29, 1940, desperately trying to follow the porter who perched the precious cargo precariously on his shoulder. Should the porter inadvertently lose his grip and the box go crashing to the floor, it would bathe the platform in Britain's most valuable military secrets—such as those listed by Bob Buderi in his book *The Invention That Changed the World*: "blueprints and circuit diagrams for rockets, explosives, superchargers, gyroscopic gun-sights, submarine detection devices, self-sealing fuel tanks, and even the early designs of the jet engine and atomic bomb " Yet, none of these technological advancements compared with the one device contained within the box Britain considered its most closely guarded secret—the resonate cavity magnetron.

A small device that fit in the palm of your hand, it resembled a clay pigeon used in skeet shooting with a few wire leads attached. Invented just eight months earlier by Birmingham University physicists J.T. Randall and H.A.H. Boot, the cavity magnetron packed a powerful punch with its ability to send out pulses of microwave radio energy on a wavelength just shy of ten centimeters, an astounding one-fifteenth the size of the standard, airborne radar transmissions then in existence. Current radar, operating with wavelengths of between ten and thirteen meters, could identify the target only within a several mile range; pinpointing the object required visual correction by the pilot. With the new cavity magnetron, however, aircraft would be able to distinguish rising U-boat periscopes from waves on the darkest of nights or identify enemy aircraft through heavy cloud cover.

Buderi put it this way in his book: "Fitted into night-fighters, such a device would generate sharper pulses in a tightly concentrated parcel of energy that would fan out

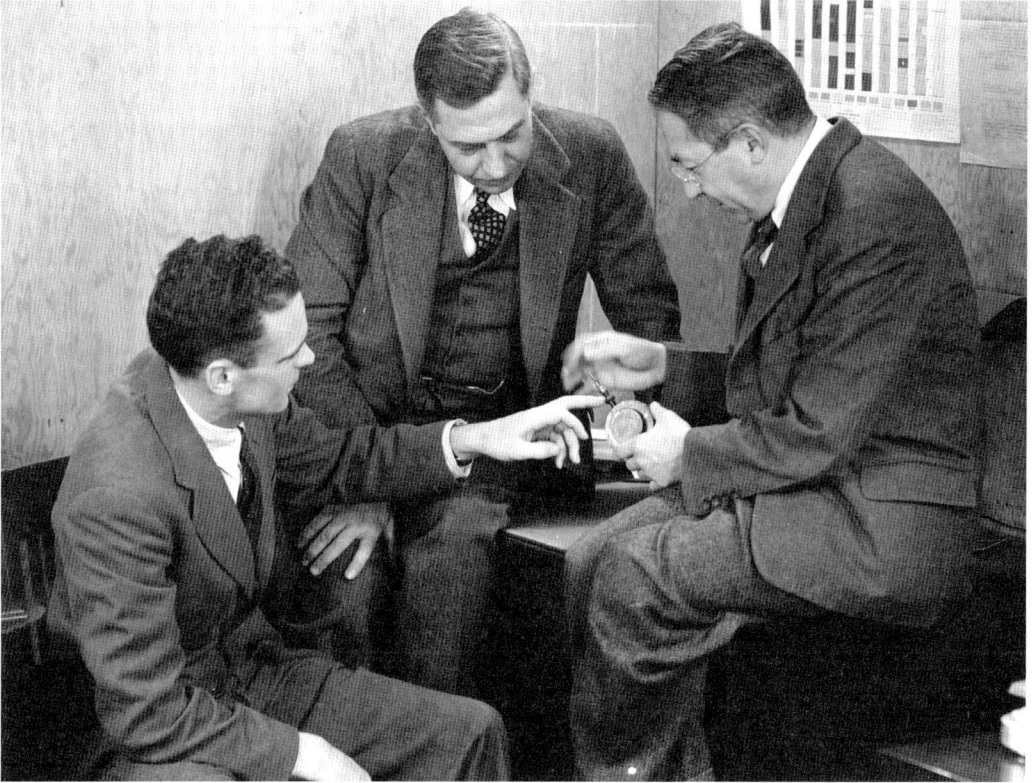

Left to Right: Edward "Taffy" Bowen, L.A. Dubridge and I.I. Rabi examine the cavity magnetron. *Photo courtesy of the MIT Museum.*

far less during the brief journey to an enemy aircraft and back, making it immensely easier for pilots to home in on their quarry even on the darkest nights."

But testing the device proved difficult. Bombing raids over London and the threat of invasion by the Germans produced power failures throughout Britain that severely hampered lab work and production capabilities. With a consistent electrical supply as well as electronics expertise from America's foremost scientists, however, those problems could be laid aside.

Bowen, the twenty-nine-year-old Welshman who ranked as one of Britain's top defense pioneers, had already developed the Chain Home network, a series of towers up to 350 feet tall that ran down Britain's south and east coasts, providing effective early radar warning of German attacks. And Bowen had developed the country's first crude airborne radar. Both of these systems were vital to Britain's defense as she guarded against Nazi armies that flanked her coastline from Norway to France and whose Luftwaffe rained bombs on London. Yet the development of these defenses would pale in comparison to the secret device Bowen now ferried as part of his covert operation.

Tapped as part of a top-secret government team with the mission to convince the Americans to assist in the final development of a revolutionary technology—one that would change the course of the war—Bowen's job was to transport the small metal solicitor's deed box from Liverpool to Washington.

The secret task took Bowen from London's Euston Station to Gladstone Dock in Liverpool. There he bordered the *Duchess of Richmond*, a Canadian liner, where he joined the others selected for this important mission. Formally called the British Technical and Scientific Mission to the United States, it was more commonly known as the Tizard Mission after its organizer, Sir Henry Tizard, rector of the Imperial College of Science and Technology and chairman of the government's key scientific committee on air defense.

The Tizard Mission comprised Bowen; John Cockcroft, Cambridge University physicist and architect of one of the world's first proton accelerators; experienced combat officers from each of Britain's three services—the Royal Air Force, the admiralty and the army; and Edgar Woodward-Nutt, an air ministry official who served as secretary.

The plan called for a full disclosure to the Americans of British technical secrets. The hope was that the United States would assist in the research, development and production of war-worthy devices. Although some powerful Brits opposed the unilateral disclosure, Prime Minister Winston Churchill directed the men on the Tizard Mission to make the offer without reciprocation.

With all aboard, the *Duchess of Richmond* pulled from the docks but soon found herself on the fringe of a bombing raid. Hunkering down for the night, she finally set sail the next morning. Upon finding the Mersey River littered with the wreckage of bombed and mined boats, the *Duchess of Richmond* had to be escorted by minesweepers. When she reached open waters, two destroyers took over. To avoid detection by German U-boats, the *Duchess* tacked her way west and sailed for Canada.

> *"When the members of the Tizard Mission brought* [a resonate cavity magnetron]
> *to America in 1940, they carried the most valuable cargo ever brought to our shores."*
> *James Phinney Baxter III, official historian of the Office of Scientific Research*
> *and Development, in The Invention That Changed the World*

At Halifax Harbor, in Nova Scotia, Canada, most of the group headed for Washington, D.C., while Bowen went to Ottawa, Ontario, to round up previously shipped equipment and meet Canadian officials who would join the exchange.

The group reconvened in Washington, D.C. There, Tizard met Vannevar Bush, a Massachusetts Institute of Technology electrical engineer who pioneered early computing. Buderi's book states: "'Of the men whose death in the summer of 1940 would have been the greatest calamity for America, the President is first and Dr. Bush would be second or third,' noted the multimillionaire investment banker, Alfred Loomis, a Bush friend destined to play a crucial role in the radar story."

The Rad Lab eventually moved to the roof where it became the main hub of activity. *Photo courtesy of the MIT Museum.*

The British and Americans met several times with much posturing from representatives on both sides of the Atlantic. The British contingency had already disclosed details of the Chain Home early warning system, Identification Friend or Foe (a radar system devised to identify friendly aircraft from foe) and a number of other radar systems, yet the men hesitated to reveal the final bargaining chip. It took several meetings before trust prevailed.

Finally, Bowen and Cockcroft pulled out the cavity magnetron. "From that moment on," Buderi writes, "things went smoothly."

THE RAD LAB

Following the Tizard Mission, construction began on a new unified research lab for the development of radar technology. Massachusetts Institute of Technology won site approval because of its history of microwave work and the fact that an additional lab on campus wouldn't raise any eyebrows.

With funding from the National Defense Research Committee, led by Bush, and support from Wall Street tycoon Alfred Loomis, sixteen scientists from the United States and the United Kingdom set secretly to work. The Radiation Laboratory (or "Rad Lab," so named to confuse the enemy) opened on November 11, 1940.

On the ground floor of its new headquarters in Building 4, with the windows painted black, twenty people gathered in a small classroom. Guards monitored the building twenty-four hours a day, seven days a week.

Work proceeded at a furious pace for the next month, and by mid-December, the lab almost doubled in size and employed thirty-six people. Outgrowing its original space, it moved upstairs to the second floor and then to the roof, where it soon became the main hub of activity.

Because the United States had not yet entered the war, financial support for the lab was tough to come by during the first year. However, things were about to change. Looming on the horizon were circumstances that would thrust the Rad Lab into the forefront of research and development, and bring unprecedented technical and financial assistance to the fledgling program.

This would set the stage for what was to come in the small town of Boca Raton, Florida.

1941 TIME CAPSULE

The United Service Organization (USO) is founded to cater to the armed forces and defense industries • A tradition begins when an organ is first played at a baseball game in Chicago • The Bulova Watch Company pays $9 for the first-ever network TV commercial • U.S. savings bonds go on sale

WORLD NEWS

• U.S. President Franklin D. Roosevelt and British Prime Minister Winston Churchill sign the Atlantic Charter.
• A massive Nazi attack is made on Russia.
• Joseph Stalin is named Soviet premier.
• Adolf Hitler initiates his "Final Solution."
• Selective Service is extended; age of service goes from 21–35 to 20–45.
• The British sink Germany's *Bismark* in retaliation for their attack on the English ship *Hood*.
• German actress Marlene Dietrich becomes a U.S. citizen.

NATIONAL NEWS

• NBC issues the first commercial television license.
• M&M's® Plain Chocolate Candies and Cheerioats (renamed Cheerios in 1946) are first introduced.
• The National Gallery of Art in Washington, D.C., opens with the bequest of Andrew W. Mellon.
• President Franklin D. Roosevelt is inaugurated for a third term.
• The Ford Motor Company signs the first contract with a labor union.
• Religious training in public schools is declared unconstitutional.
• Japanese kill more than two thousand seamen at Pearl Harbor, Hawaii.

SPORTS

• World Series champion—New York Yankees • U.S. Open golf champion—Craig Wood • Pro football champion—Chicago Bears • Indianapolis 500 winner—Floyd Davis & Mauri Rose • Stanley Cup winner—Boston Bruins • NCAA basketball champion—Wisconsin • College football champion—Minnesota • Heisman Trophy winner—Bruce Smith from Minnesota

MUSIC

• "Chattanooga Choo Choo"—Glenn Miller
• "Dream Valley"—Sammy Kaye
• "Green Eyes"—Jimmy Dorsey
• "Amapola"—Jimmy Dorsey
• "Daddy"—Sammy Kaye
• "Down Argentina Way"—Bob Crosby
• "Elmer's Tune"—Glenn Miller
• "High on a Windy Hill"—Jimmy Dorsey
• "Intermezzo"—Guy Lombardo
• "Till Reveille"—Kay Kyserv

MOVIES

• *How Green Was My Valley*—Academy Award winner
• *Citizen Kane*
• *Dumbo*
• *Sergeant York*
• *Meet John Doe*
• *The Maltese Falcon*

COST OF LIVING

New house	$4,075.00
Average income	$1,777.00 per year
New car	$850.00
Average rent	$32.00 per month
Tuition at Harvard University	$420.00 per year
Movie ticket	$.30 each
Gasoline	$.13 per gallon
U.S. postage stamp	$.03 each

Food

Granulated sugar	$.59 for 10 pounds
Vitamin D milk	$.54 per gallon
Ground coffee	$.45 per pound
Eggs	$.20 per dozen
Fresh baked bread	$.08 per loaf

1941

THE RAD LAB

While the outside world prepared to celebrate Christmas, Wall Street tycoon Alfred Loomis scrawled on the blackboard an ambitions schedule of target dates for the airborne interception (AI) radar system:

- *January 6, 1941—a microwave system working on the Roof Lab.*
- *February 1, 1941–a working system mounted in a B-18 bomber supplied by the army.*
- *March 1, 1941–a working system adapted for an A-20A attack bomber*

On January 4, just eight weeks after opening, the Rad Lab received its first echoes by aiming a radar beam at the Christian Science church tower in Boston, Massachusetts. Premature jubilation gave rise to sober reality however, as scientists discovered there were as many things wrong with the system as there were right. The device proved too big and unwieldy, and had two antennas for transmitting and receiving—instead of just one. The team went back to the drawing board.

January 10 marked the birth of the single-dish system, but with frequent breakdowns, deadlines came and went. Finally, on March 27, the lab ran a test in a B-18 over Cape Cod. On that run, researchers discovered that the device could not only pick up planes at a distance of three miles but also detect submarines at a five-mile range. From then on, the Rad Lab added air-to-surface vessels to its list of research and development needs.

The Rad Lab soon required more space and money. MIT allocated additional research space while John D. Rockefeller supplied the financial support. In the spring of 1941, the Rad Lab staff grew to more than 140: 90 physicists and engineers; 45 mechanics, technicians, guards and secretaries; and 6 Canadian guest scientists.

The SCR-584, an automatic tracking radar, protected ground troops from air attacks. *Photo courtesy of the Historical Electronics Museum.*

In November, at Fort Hancock, New Jersey, the Army Signal Corps saw the first prototype of the SCR-584 automatic tracking radar, one of the most important radar sets produced by the Rad Lab. Throughout World War II, thousands of SCR-584s would be used by the army to protect ground troops from air attacks.

ENTER BOCA RATON

Far to the south of the Rad Lab, sandwiched between the popular east-coast Florida resort towns of Palm Beach and Miami, lay the quiet community of Boca Raton. But quiet would soon give way to the roar of war.

Boca Raton—whose name suggests a number of historic interpretations that range from "rat or mouse mouth," to "hidden rocks" or "thieves' inlet"—was a community of 723 residents. There, nothing much went on except the sunshine and azure waves that lapped at its sandy beaches.

The town consisted of a small airport, town hall, two general stores (Max Hutkin's and Tony Brenk's), two gas stations, a roadside restaurant, two taverns (Zim's and Brown's),

Covered by camouflage netting and surrounded by a sandbag bunker, the SCR-584 is readied for action. *Photo courtesy of the Historical Electronics Museum.*

a bus terminal, a railroad depot and the most important commercial enterprise of them all: the private Boca Raton Club.

While Boca's population hovered near seven hundred for most of the year, in season (from November to April), the population swelled to close to three thousand as tourists and their domestic help made their way to the Boca Raton Club or to the private residences on or close to the beach. There, the tourists would escape cold dreary Northern winters to enjoy mild temperatures and frolic in the warm sun and frothy surf. Those who stayed at the Boca Raton Club wallowed in luxury.

The town evolved from humble beginnings in 1895 with the arrival of its first settler, Captain Thomas Moore Rickards. An engineer and landowner, he cultivated the sandy soil into pineapple plantations, orange groves and vegetable farms in the early 1900s. Henry M. Flagler's railroad made its debut in Boca Raton at about the same time with steel tracks strung from West Palm Beach to Miami, allowing access to the mostly wooded land in Boca Raton. Even a colony of Japanese farmers, the Yamato Colony (pronounced YAH-muh-tow), settled in Boca Raton, bringing with them strong backs and work ethics as they labored under difficult conditions to clear virgin hammocks of pine and palmettos.

During the early 1940s, Boca Raton had a population of 723. This is a view of the town looking south along Federal Highway. *Photo courtesy of the Boca Raton Historical Society.*

Captain Thomas Moore Rickards (fourth from right) and survey crew stop in the shade of Flagler's FEC Railroad. *Photo courtesy of the Boca Raton Historical Society.*

The colony of about forty Japanese proved successful in cultivating seventy acres of pineapple and one hundred acres of citrus and vegetables until 1908 when blight wiped out the pineapples and competition from Cuba destroyed the domestic market. Many of the colonists went back to Japan; however, a few families remained, including Hideo Kobayashi, who owned five hundred acres. Later, the U.S. government would confiscate part of his land to make way for the Boca Raton Army Air Field.

The land boom of the 1920s brought Addison Mizner to Boca Raton with the dream of building a sixteen-thousand-acre residential/resort community. His vision produced twenty-nine resort homes in the Old Floresta and Spanish Village sections of the town as well as the Cloister Inn, a luxury resort of one hundred rooms that opened with much fanfare February 6, 1926. But in July of that year, 117 Florida banks went belly up, ending Mizner's dream. Clarence H. Geist, one of Mizner's backers, believed in his vision and, in 1927, acquired $7 million in Mizner debt and paid $71,000 for the inn.

In the late 1920s Geist enlarged the inn, renaming it the Boca Raton Club. Additions included three hundred guest rooms within a six-story building as well as a 350-foot-square entrance courtyard and a saltwater pool. Guests motored among stately palms that lined the drive from Camino Real to the front doors of the club or arrived by private yacht at the club's docks.

The Kamiya children sit on the front porch of their home. The family was one of several Japanese farmers who became part of the Yamato Colony. *Photo courtesy of the Boca Raton Historical Society.*

Mizner Development Corporation advertises club memberships and properties. *Photo courtesy of the Boca Raton Historical Society.*

Clarence and Florence Geist arrive at the Boca Raton depot. *Photo courtesy of the Boca Raton Historical Society.*

View of the Boca Raton Club from Lake Boca Raton. *Photo courtesy of the Boca Raton Historical Society.*

Other amenities included tennis courts, trap shooting, horseback riding and two healthatoriums, one each for the ladies and men. Also provided were a children's dining room, separate dining facilities for chauffeurs and maids, a garage, a barbershop, a beauty parlor and a number of boutiques.

Geist ran the club for the next eight years as a private club with only members and their guests admitted through guarded gates. After members paid an initiation fee of $5,000, which gave them a proprietary interest in the club, they could enjoy the unprecedented amenities for dues of a mere $100 per year.

Little did Geist know that in a few short years, his labor of love would soon become a cog in the great war machine.

During the late 1920s, August H. Butts, a farmer from Lake County in central Florida, bought up land vacated during the land bust, and in the 1930s, he became the largest landowner in the area. Using the most up-to-date farming methods and irrigation systems, the farm became the principal producer of green beans in Florida. It eventually covered 3,500 acres (six square miles) in the Boca Raton area and employed more than four hundred local and migrant workers in season.

In 1933, the Florida Legislature named A.B. McMullen, a pioneer Florida flier, as director of an aviation division within the State Road Department. His job—developing the air industry in Florida—included a ten-year plan in which he targeted Boca Raton as an area in need of an airport. Donald W. Curl, author of

Palm Beach County, explains McMullen's rationale for selecting Boca Raton: "The terrain afforded no natural areas for emergency landings, which were fairly common for the planes of that era."

At the same time, Geist insisted that many of his club guests, who now owned airplanes, needed a convenient way to travel to and from Boca Raton. Not only did he want an airstrip, he wanted a hangar and terminal as well. Recognizing an opportunity to combine construction of the airport with the federal government's emergency relief activities, McMullen urged Geist to ask the town to build an airport.

Gordon B. Anderson, Boca Raton Club's general manager, organized and planned the airport project. Neither the Town of Boca Raton nor the State of Florida was able to finance such a major undertaking, but the Model Land Company that, according to Curl, "held the land grants given to Henry M. Flagler to build the Florida East Coast Railway," made available land decried worthless for agricultural purposes. The town contributed $11,869 for the project, and the New Deal Works Progress Administration (WPA) granted $38,537.

Construction began in late 1936 with grass runways completed in 1937. The hangar and terminal never got off the ground.

When Geist died in 1938, the club reverted to a trust with Geist's wife, Florence, acting as the club's manager. Yet Geist's foresight in enlarging the Boca Raton Club and

Lobby of the Boca Raton Club. *Photo courtesy of the Boca Raton Historical Society.*

Former Boca Raton Mayor Fred Aiken examines beans at a roadside stand on the Butts farm. *Photo courtesy of the Boca Raton Historical Society.*

Clarence Geist's aerial photographer lands his plane in a field west of Boca Raton in 1929. *Photo courtesy of the Boca Raton Historical Society.*

spearheading efforts to obtain an airport would prove pivotal in Boca Raton's ability to attract a military installation to its tranquil community.

Young Men Join the War Effort

While Boca Raton went about its quiet business, war raged across Europe and the Pacific was heating up. Kenny Davey, like tens of thousands of other young men across the county, stepped forward to enlist in the armed services.

Kenny Davey, 26, stood at attention—stomach in, chest out. In his boxer shorts, and not much else, his mind swirled with a mixture of anxiety and anticipation. This was his last chance. If he failed to make the weight restrictions this time, he'd be drafted into the regular army, and his desire to become an Army Air Corps pilot would vanish as quickly as a dream.

Flight surgeon Dr. Henry eyed the candidate before him with skepticism. In the last seventy days, although Davey proved to be a specimen of good health, he had failed the weight requirement twice. To qualify for pilot training, Davey's five-foot-six-inch frame needed to weigh 158 pounds, but the scale registered 178.

A solid young man, Davey grew up on a working farm in southern Minnesota. The youngest of thirteen children, he plowed cornfields, bailed hay, slaughtered hogs, milked cows by hand and finished high school, the only one of his family to do so.

Working and going to college for the last six years at Washington University in St. Louis, Missouri, Davey finally had graduation in his sights. But at the last minute, he learned he got a D on a psychology final. This disappointment put him one point away from his degree. Yet that one point may as well have been a thousand; before he could retake the class, war broke out and Davey faced the draft. Returning to school now seemed a lifetime away.

Given just ten days to shed twenty pounds before reporting for his third and final physical exam, Davey adhered to his own strict weight-loss program. He wore wool clothes as he worked outside in the summer heat on the farm and took saunas in the town barbershop. He knew it all came down to this moment—8:00 a.m. Wednesday morning in the federal building at Omaha, Nebraska.

Dr. Henry examined the record before him, then peered over his reading glasses at Davey. "Step on the scale," the doctor said, pointing at it with the end of his pencil. Davey dutifully complied.

Slowly sliding the weights along the notched metal shaft, Dr. Henry stopped as the pointer registered the balance. He then shook his head. "Is that the best you can do? Sixteen pounds in ten days?"

Davey swallowed hard. "Yes, sir," he managed.

Dr. Henry took one more look at the file then a final look at Kenny. "The only way you can ever reduce your weight to 158 pounds is with an eraser," he said.

Aerial view of the Boca Raton Airport in 1940. *Photo courtesy of the Boca Raton Historical Society.*

Thanks to Dr. Henry and his eraser, Ken Davey was about to begin his life as an Army Air Corps cadet. On July 27, 1941 at Fort Snelling in St. Paul, Minnesota, he took an oath "to defend the constitution of the United States from enemies foreign and domestic."

With thousands of others during World War II, Davey would become the backbone of our nation's air defense. In planes he would fly, airborne radar devices would be installed and radar operators, trained at the Boca Raton Army Air Field would become an integral part of winning the war.

Members of the Puerto Rico Rum Company land at the Boca Raton Airport. *Photo courtesy of the Boca Raton Historical Society.*

Town Council Visited by Navy

In 1941, Jones Cleveland (J.C.) Mitchell was in his third year as mayor of Boca Raton. Having come to the area in 1923, he invested in real estate and worked his way up through community channels to become chairman of the town council and subsequently mayor.

A savvy politician, Mitchell read the political temperature of the times and realized war was eminent. There would soon be a need to build military bases in strategic locations across the country, including South Florida. With his political contacts, influence and vision, and Geist's foresight regarding the building of the airfield and expansion of the Boca Raton Club, Mitchell felt the town was now poised on the brink of something big—something he wasn't about to let slip through his fingers.

Persuading the town council to support him, Mitchell packed his bags and headed for Washington, D.C., with the hope of convincing officials that Boca Raton would be

Like many young men, Kenny Davey chose to join the Army Air Corps rather than be drafted into the regular Army. *Photo courtesy of Colonel Kenneth W. Davey USAF (Ret.).*

Mayor J.C. Mitchell was instrumental in bringing the Boca Raton Army Air Field to Boca Raton. *Photo courtesy of the Boca Raton Historical Society.*

the ideal location for one of these new bases. When he arrived, however, he went head-to-head against a throng of mayors from towns and cities across the country—all with the same agenda, all jockeying for position.

In *Boca Raton, A Pictorial History*, authors Donald W. Curl and John P. Johnson described J.C. Mitchell's Washington experience: "Shuffled from the War Department to the Navy and to the Army Air Corps, Mitchell finally received a promise that Boca Raton would at least be considered as an air base site."

That promise came in the form of a visit from Major General Henry "Hap" Arnold, commanding general of the Army Air Force and architect of America's emerging, independent air force. Sometime in 1941, Arnold made an inspection swing through Florida, stopping at Morrison Field in West Palm Beach. Just before his arrival, a tropical storm pushed through, leaving six inches of standing water. Jackie Waldeck captured Arnold's response in her article "How Boca Helped Win the War," published in the March/April 1982 edition of *Boca Raton Magazine*: "If Boca Raton is like this," he snapped, "we're not interested!"

According to Waldeck, Mayor Mitchell and a local delegation were on hand and assured the general that the airport at Boca Raton remained dry. When Arnold arrived and saw this was true, he instructed Colonel Thomas L. Bryan to commence a more in-depth inspection.

Colonel Bryan had recently served an extensive tour in England, where he studied the role communications played in global warfare. It culminated in the initiation of discussions, plans, studies and experiments about various phases of communication used in theaters of war. As a result, he created a technical-training school for instruction in radar.

About this same time, the navy also expressed interest in Boca Raton, and on November 5, 1941 at 2:30 p.m., the town council called a special meeting. T.D. Giles presided and called upon Mayor J.C. Mitchell, who introduced Lieutenant Commander D.M. Campbell and Lieutenant C.M. Ewan of the U.S. Navy Reserves, as well as Captain K.E. Voelter of the U.S. Marine Corps Reserves. The purpose of the meeting was to discuss the navy's proposal that the Boca Raton airport be secured for naval training purposes for five years at a cost of $1 per year.

The navy would improve the airfield, including mowing, seeding and rolling; filling in low spots; and performing other necessary work. In addition, naval officials desired to acquire land southwest of the airport so the airfield could be approximately square. Navy plans also included a portable operations building and a well.

Both sides agreed planes would avoid flying at low altitudes over the town and nearby congested areas. Private or commercial airplane traffic could continue off the field. This traffic consisted of a few airplanes owned by wealthy individuals who spent time in Boca Raton during the winter months.

The council adjourned at 2:45 p.m. to the airport to witness the planes in action, then reconvened in council chambers at 3:15 p.m. where the resolution was unanimously adopted.

Honolulu Star-Bulletin 1st EXTRA

HONOLULU, TERRITORY OF HAWAII, U. S. A., SUNDAY, DECEMBER 7, 1941 ★ PRICE FIVE CENT

WAR!
OAHU BOMBED BY JAPANESE PLANES

(Associated Press by Transpacific Telephone)

SAN FRANCISCO, Dec. 7.—President Roosevelt announced this morning that Japanese planes had attacked Manila and Pearl Harbor

SIX KNOWN DEAD, 21 INJURED, AT EMERGENCY HOSPITAL

Attack Made On Island's Defense Areas

By UNITED PRESS

WASHINGTON, Dec. 7.—Text of a White House announcement detailing the attack on the Hawaiian islands is:

"The Japanese attacked Pearl Harbor from the air and all naval and military activities on the island of Oahu, principal American base in the Hawaiian islands."

Oahu was attacked at 7:55 this morning by Japanese planes.

The Rising Sun, emblem of Japan, was seen on plane wing tips.

Wave after wave of bombers streamed through the clouded morning sky from the southwest and flung their missiles on a city resting in peaceful Sabbath calm.

According to an unconfirmed report received at the governor's office, the Japanese force that attacked Oahu reached island waters aboard two small airplane carriers.

It was also reported that at the governor's office either an attempt had been made to bomb the USS Lexington, or that it had been bombed.

CITY IN UPROAR

Within 10 minutes the city was in an uproar. As bombs fell in many parts of the city, and in defense areas the defenders of the islands went into quick action.

Army Intelligence officers at Ft. Shafter announced officially shortly after 9 a. m. the fact of the bombardment by an enemy but long previous army and navy had taken immediate measures in defense.

"Oahu is under a sporadic air raid," the announcement said.

"Civilians are ordered to stay off the streets until further notice."

CIVILIANS ORDERED OFF STREETS

The army has ordered that all civilians stay off the streets and highways and not use telephones.

Evidence that the Japanese attack has registered some hits was shown by three billowing pillars of smoke in the Pearl Harbor and Hickam field area.

All navy personnel and civilian defense workers, with the exception of women, have been ordered to duty at Pearl Harbor.

The Pearl Harbor highway was immediately a mass of racing cars.

A trickling stream of injured people began pouring into the city emergency hospital a few minutes after the bombardment started.

Thousands of telephone calls almost swamped the Mutual Telephone Co., which put extra operators on duty.

At The Star-Bulletin office the phone calls deluged the single operator and it was impossible to handle the flood of calls. Here also an emergency operator was called.

HOUR OF ATTACK—7:55 A. M.

An official army report from department headquarters, made public shortly before 11, is that the first attack was at 7:55 a. m.

Witnesses said they saw at least 50 airplanes over Pearl Harbor.

The attack centered in the Pearl Harbor. Army authorities said:

"The rising sun was seen on the wing tips of the airplanes."

Although martial law has not been declared officially, the city of Honolulu was operating under M-Day conditions.

It is reliably reported that enemy objectives under attack were Wheeler field, Hickam field, Kaneohe bay and naval air station and Pearl Harbor.

Some enemy planes were reported shot down.

The body of the pilot was seen in a plane burning at Wahiawa.

Oahu appeared to be taking calmly after the first impact of surprise

ANTIAIRCRAFT GUNS IN ACTION

First indication of the raid came shortly before 8 this morning when antiaircraft guns around Pearl Habor began sending up a thunderous barrage.

At the same time a vast cloud of black smoke arose from the naval base and also from Hickam field where flames could be seen.

BOMB NEAR GOVERNOR'S MANSION

Shortly before 9:30 a bomb fell near Washington Place, the residence of the governor. Governor Poindexter and Secretary Charles M. Hite were there.

It was reported that the bomb killed an unidentified Chinese man across the street in front of the Schuman Carriage Co. where windows were broken.

C. E. Daniels, a welder, found a fragment of shell or bomb at South and Queen Sts. which he brought into the City Hall. This fragment weighed about a pound.

At 10:05 a. m. today Governor Poindexter telephoned to The Star-Bulletin announcing he has declared a state of emergency for the entire territory.

He announced that Edouard L. Doty, executive secretary of t he major disaster council, has been appointed director under the M-Day law's provisions.

Governor Poindexter urged all residents of Honolulu to remain off the street, and the people of the territory to remain calm.

Mr. Doty reported that all major disaster council wardens and medical units were on duty within a half hour of the time the alarm was given.

Workers employed at Pearl Harbor were ordered at 10:10 a. m. not to report at Pearl Harbor.

The mayor's major disaster council was to meet at the city hall at about 10:30 this morning.

At least two Japanese planes were reported at Hawaiian department headquarters to have been shot down.

One of the planes was shot down at Ft. Kamehameha and the other back of the Wa-

Hundreds See City Bombed

Hundreds of Honolulans who hurried to the top of Punchbowl soon after bombs began to fall, saw spread out before them the whole panorama of surprise attack and defense.

Far off over Pearl Harbor the white sky was polka dotted with anti-aircraft smoke.

Rolling away from the navy base were billowing clouds of ugly black smoke. Sometimes a burst of flame reddened the black centers of the smoke.

Out from the silver-surfaced mouth of the harbor a flotilla of destroyers streamed to battle, smoke pouring from their stacks.

Schools Closed

All schools on Oahu, both public and private, will remain closed until further notice. Edward L. Doty, territorial director of civilian defense, announced at 11 a. m. today. This does not apply elsewhere in the territory.

Names of Dead and Injured

The city emergency hospital reported at 10:30 a list of 6 killed and 21 injured.

The complete list will be carried later. Here is a partial list:

Frau Lopes, 34, of 3641 Kaimuki Rd St., was reported at 9:30 a. m to be in serious condition from wounds in the upper abdomen.

Bernice Guevara, 12, 1706 Kalihi St., is suffering from a shrapnel wound, lacerations on the right leg and left arm.

A Portuguese girl, unidentified, 16 years old, died on arrival from puncture wounds.

Another victim who died on arrival was Frank Ohashi, 38, 710 Kamananui St., from puncture wounds in the chest.

Cecelia Broadly, 38, Momoma gardens, was released from the hospital after treatment for lacerations.

Three were reported injured and one reported killed from the bomb that fell at Fort and School Sts.

Editorial

HAWAII MEETS THE CRISIS

Honolulu and Hawaii will meet the emergency of war today as Honolulu and Hawaii have met emergencies in the past—coolly, calmly and with immediate and complete support of the officials, officers and troops who are in charge.

Governor Poindexter and the army and navy leaders have called upon the public to remain calm; for civilians who have no essential business on the streets to stay off; and for every man and woman to do his duty.

That request, coupled with the measures promptly taken to meet the situation that has suddenly and terribly developed, will be heeded.

Hawaii will do its part—as a loyal American territory.

In this crisis, every difference of race, creed and color will be submerged in the one desire and determination to play the part that Americans always play in crisis.

BULLETIN

Additional Star-Bulletin extras today will cover the latest developments in this war.

The front page of the afternoon edition of the December 7, 1941 *Honolulu Star* denounces the earlier attack on Pearl Harbor and Manila. *Photo courtesy of the Boca Raton Historical Society.*

Air Corps Technical School of Radar Established

On November 29, 1941, Colonel Bryan directed the temporary establishment of the Army Air Corps Technical School of Radar at Scott Field, Illinois. The school would be dedicated to the training of troops in the operation and repair of airborne radar. Then came December 7 and the attack on Pearl Harbor—an event that forced America headlong into the war.

At the Rad Lab in Massachusetts, the attack brought financial concerns to a screeching halt. Money poured in, accelerating radar research and development. In turn, this meant new radar operators and mechanics would enter military service, needing to be trained, housed and fed. Space at Scott Field was limited, and it soon became apparent facilities needed to be expanded. The school also needed a permanent home—a place close to water for target practice.

Sites at the Great Lakes were considered but rejected because of the frozen lakes during the winter months. A more suitable site would need to have moderate temperatures and possess good flying weather. The Signal Corps radar school, a ground radar-training facility, had already been established at Camp Murphy—located in Hobe Sound, Florida, on the Loxahatchee River, just north of Palm Beach. Top military brass wanted to locate the new base nearby.

It was now time to ferret out a site.

1942 Time Capsule

A nightly "dim-out" begins along the East Coast • The Women's Army Auxiliary Corps begins first basic training at Fort Des Moines, Iowa • "Rosie the Riveter" becomes a national symbol for women entering the workforce • The United States begins rationing coffee

World News

• The United States and Philippines give up Bataan; thirty-six thousand troops are led on the infamous Bataan Death March.
• The British begin heavy raids on German cities.
• The Battle of Midway, from June 3 to 6, ends with the first major Japanese loss.
• Major General James H. Doolittle leads a bombing group over Tokyo.
• General Francisco Franco ousts cabinet members in Madrid, Spain, obtaining full control of government.
• Holland's Queen Wilhelmina pays a state visit to the White House.

National News

• The Office of War Information is created by President Franklin D. Roosevelt.
• War bonds are introduced, raising $13 billion by December 23.
• Gasoline rationing goes into effect in the United States.
• The Office of Civil Defense is established.
• The Manhattan Project begins.
• Zoot suits are a fashion statement for many men.
• Kellogg's Raisin Bran® and instant coffee are introduced.
• The draft age is lowered to 18 years of age.

Sports

• World Series champion—St. Louis Cardinals • U.S. Open golf champion—Was not held in 1942 • Pro football champion—Washington Redskins • Indianapolis 500 winner—Was not held in 1942 • Stanley Cup winner—Toronto Maple Leafs • NCAA basketball champion—Stanford • College football champion—Ohio State • Heisman Trophy winner—Frank Sinkwich from Georgia

Music

• "Jingle, Jangle, Jingle"—Kay Kyser
• "Jersey Bounce"—Benny Goodman
• "White Christmas"—Bing Crosby
• "Blues in the Night"—Woody Herman
• "Deep in the Heart of Texas"—Alvino Ray
• "He Wears a Pair of Silver Wings"—Kay Kyser
• "I've Got a Gal in Kalamazoo"—Glenn Miller
• "One Dozen Roses"—Dick Jurgens
• "Remember Pearl Harbor"—Sammy Kaye
• "Sleepy Lagoon"—Harry James

Movies

• *Mrs. Miniver*—Academy Award winner
• *Bambi*
• *Saboteur*
• *Now, Voyager*
• *To Be or Not to Be*

Cost of Living

New house	$3,775.00	
Average income	$1,885.00 per year	
New car	$920.00	
Average rent	$35.00 per month	
Tuition at Harvard University	$420.00 per year	
Movie ticket	$.30 each	
Gasoline	$.15 per gallon	
U.S. postage stamp	$.03 each	

Food

Granulated sugar	$.59 for 10 pounds
Vitamin D milk	$.60 per gallon
Ground coffee	$.45 per pound
Eggs	$.20 per dozen
Fresh baked bread	$.09 per loaf

1942

A PIVOTAL YEAR

In January 1942, Manila and the Philippines fell to Japan; in February, the Japanese defeated the Allied strike force at Java and the Netherland Indies. On Bataan, in the Philippines, 76,000 U.S. and Filipino troops, low on food and ammunition, surrendered to the Japanese and were forced to march to distant camps. In route, 5,200 Americans died in what became known as the Bataan Death March.

With the war in full swing in the Atlantic and the Pacific, further development of radar became a priority at the Rad Lab in Massachusetts. Simultaneously, military priorities included construction of new bases to house and train the hordes of young men now entering the service. Many would be funneled into training for radar operations and repair.

In her article "World War II in Boca Raton: The Homefront," published in the Fall 1985 edition of the *Spanish River Papers,* Drolleen P. Brown noted that top military brass considered Boca Raton for its "natural environment, good flying weather throughout the year, proximity to the ocean, and, for the purpose of security, a location not easily accessible to the enemy."

Yet the enemy was closer than anyone could have imagined. It lurked just off the East Coast under cover of the blue Atlantic.

WAR COMES TO FLORIDA

Using U-boats to torpedo the vessels bringing life-saving fuel and supplies to Great Britain, the Nazis put a stranglehold on the island nation. Without these crucial supplies, England would not be able to continue its resistance. And if she fell, the Germans had a clear avenue to transport their massive war machine to the United States.

Gulfland burns in the water off Hobe Sound, Florida. In the first six months of Operation Drumbeat, the Germans sank 397 ships and killed close to five thousand people. *Photo courtesy of Florida State Archives.*

With much of the U.S. Navy sunk at Pearl Harbor and the sights of other commissioned vessels turned toward Japan, Adolf Hitler's submarines launched Operation Drumbeat, a U-boat assault targeting commercial vessels traveling the East Coast shipping lanes of the United States.

In Jackie Waldeck's article, "How Boca Helped Win the War," she wrote: "Operating underwater, they [the U-boats] could not be seen and little defensive action could be taken against them. Furthermore, most merchant ships were unarmed, and the few that possessed armament were outgunned by the German submarines."

In his book *War in Paradise*, Eliot Kleinberg explained Hitler's mindset: "The German strategy was to interrupt the flow of supplies along the U.S. coast and to England, lay waste to the Allies' merchant fleet and strike a propaganda blow by letting Americans watch burning ships from their beaches."

And so they did.

"The Battle of Florida," by Philip Wylie and Laurence Schwab (*Saturday Evening Post*, March 11, 1944), stated:

> By night one could see sudden pillars of fire at sea and by day, dense clouds of smoke. In one instance, a torpedo missed its target and plowed toward land. It struck a sandbar a

Army wives and civilians staffed the Boca Raton spotting tower, which was operated by the military. *Photo courtesy of the Boca Raton Historical Society.*

hundred feet off the beach of a resort town. The blast broke the windows in the hotels along the ocean front for a half mile. Scores of tankers exploded, burned, keeled over, and sank within sight of land. Indeed a man with an ordinary rifle could have entered the war against the Germans by firing shots from the Florida beaches.

At that time American defense facilities from Key West to Palm Beach consisted largely of some obsolete training planes, harbor patrol boats, lighthouse tenders, and many private yachts. The bewildered citizens wondered where its own Navy was. The Japanese were spreading in the Pacific; the Navy was there. England and Russia were both beleaguered and were depending upon the long supply line from America; the Navy was there protecting it. Nothing was left for the defense of our coastal shipping. Any coastal town in southern Florida could have been shelled from the ocean.

At first the Navy Department released detailed accounts of the torpedoed vessels. Censorship soon followed. But thousands stood on the beaches and watched the war at sea.

On February 19, 1942, the first victim, 456-foot-long *Pan Massachusetts,* took two torpedoes from U-128 twenty miles east of Cape Canaveral. Built in 1918, the vessel—an 8,200-ton oil tanker with a crew of thirty-eight—was heading north from Texas to New York when the attack occurred. A British tanker picked up survivors, but twenty of the crew perished.

Martha, 10, and Peter Barrett ,7, were spotters in the tower. They looked for German submarines and planes. *Photo courtesy of Dr. Peter Barrett.*

Rubber planes hung from the ceiling to assist spotters in recognizing aircraft. This model of a U.S. Army A-24 is 6½ x 5 inches. On the left wing base is printed the following information: "US Navy, SBD, July 42." *Photo courtesy of Dr. Peter Barrett.*

February 21 and 22 saw a repeat performance as three more tankers, *Republic, Cities Services Empire* and *W. D. Anderson*, succumbed to Hitler's underwater barrage. Silhouetted against lights from cities and tourist-laden oceanfront hotels, the ships were literally "sitting ducks in a barrel."

From sandy beaches, spotting towers rose every three miles up and down the East Coast. Under the jurisdiction of the military, dedicated civilians manned the towers twenty-four hours a day, seven days a week. As "spotters," the civilians were to watch for German submarines and aircraft, and report anything suspicious.

One of the spotters in the Boca Raton tower was seven-year-old Peter Barrett. He lived with his family in a four-villa complex known as the Boca Raton Villas. Built by his grandfather, Edward E. Barrett, the villas sat on the ocean just across from Lake Boca Raton and the Boca Raton Club. Peter, his mother, Jessie, and ten-year-old sister, Martha, often took turns in the tower.

"The watch tower was located several miles north of Palmetto Park Road, stood about thirty feet high and as was painted a dreary Navy gray color," Peter Barrett recalled. "The platform held an enclosed office area with windows, and there was a walkway around the platform. Inside the office were two phones, one of which was red in color. We were told in no uncertain terms that we should never touch the red phone unless we saw something very important."

"Study materials, showing the silhouettes of submarines and aircraft, were scattered about the table, photos and diagrams were stuck to the wall, and black, hard rubber models of aircraft hung from the ceiling. The war was very real to us, and within a short time I knew every airplane in the world," said Barrett.

While the Barretts never saw anything suspicious during their watch, they heard of German subs being spotted from their tower during the course of the war. Later, they would experience firsthand the effects of German landings—right in their own back yard.

As another deterrent for the U-boats, blackouts were imposed along the coast from dusk to dawn. Still, the Germans persisted. Tankers sank.

At Morrison Field in West Palm Beach, the Coastal Picket Patrol and the Mosquito Fleet, a Civil Air Patrol flotilla of pleasure and charter boats, patrolled for subs and rescued survivors. The burned and charred bodies of those not lucky enough to survive the infernos at sea floated to shore. Oil coated the beaches.

With Hitler's persistence, Operation Drumbeat was about to bring the war to Boca Raton's front door.

On May 8 at 12:13 p.m., several miles off Boca Raton, U-564 attacked the *Ohioan*, carrying 6,000 tons of ore, 1,300 tons of licorice root and 300 tons of wool. It sank in two minutes in 550 feet of water. Fifteen of the thirty-seven crewmen lost their lives. The survivors swam to lifeboats torn from the ship when it sank and were later picked up and taken to West Palm Beach.

Edward E. Barrett, Peter's grandfather, wrote from his ocean side home in Boca Raton to the *LaGrange Citizen* (in LaGrange, Illinois) and told how his friends from

Edward E. Barrett (center) sits at the beach with sons Hollis J. Barrett (left) and Edward D. Barrett (right). He owned the Boca Raton Villas, four individual oceanfront units he built in 1939. *Photo courtesy of Dr. Peter Barrett.*

New York, who spent the winter in Florida's Fort Lauderdale, stopped by his home that morning. After saying goodbye, they drove north along the ocean, admiring a big freighter in the Gulf Stream. Suddenly, it stood on end and sank, stern first.

"The shock of seeing the ship torpedoed before my friends' eyes was so great as to halt their northern trip for two hours at Delray Beach [in Florida], and to put one of the ladies in the hospital in New York for a week after they reached home," Edward E. Barrett wrote.

The very next day, three-and-a-half miles off Delray Beach, U-564 made its presence known by striking the *Lubrafol*. Carrying 2.5 million gallons of fuel oil, the torpedo set the tanker ablaze and sent thirteen crew members to their watery graves.

In the first six months of Operation Drumbeat, the Germans sank 397 ships and killed close to five thousand people. Between February and May, 24 ships went down off the Florida coast, 16 within a 150-mile stretch from Cape Canaveral to Boca Raton.

German Spies in Boca Raton

One summer night while Peter Barrett slept peacefully in his bed at the Boca Raton Villas, he was suddenly awakened by a "loud and insistent banging" on the front door. With Peter's father, Hollis Barrett, in Chicago, Illinois, on business, his mother, Jessie

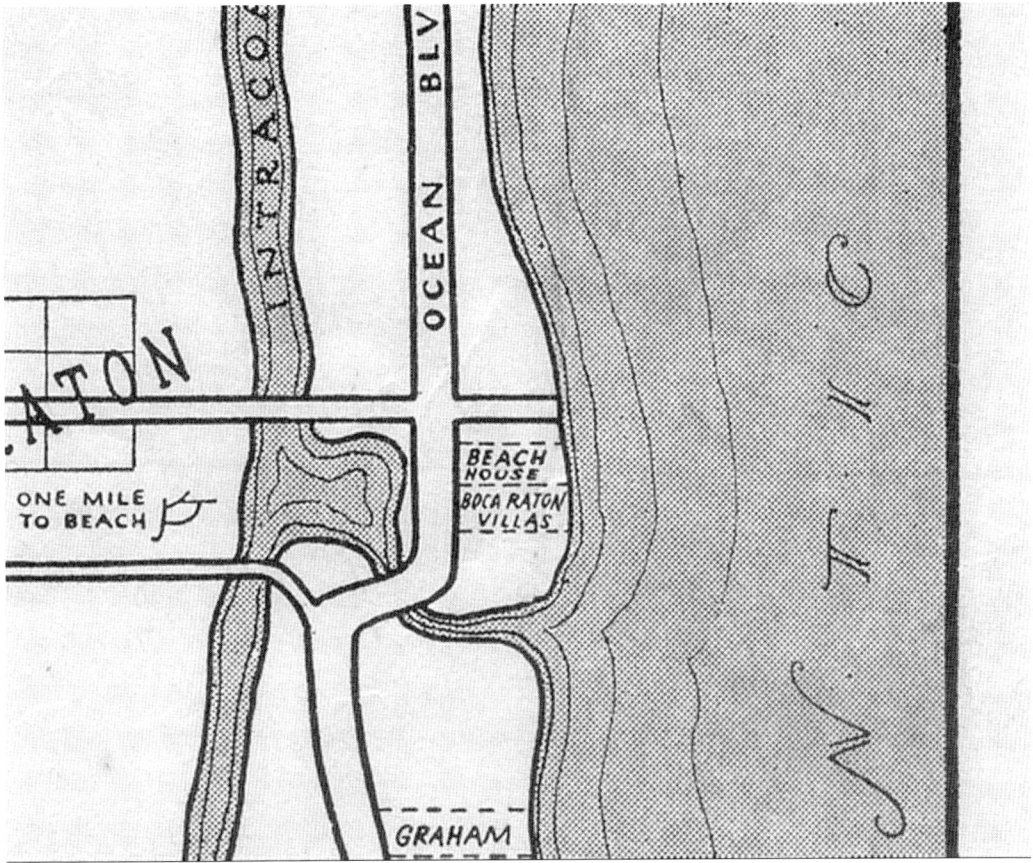

The Boca Raton Villas sat 100 yards to the south of Mrs. Dixon's Beach House Inn. Dr. Sanborn's large, walled home was immediately to the south of the villas. During the war, many mysterious happenings occurred in and around these dwellings. *Photo courtesy of the Boca Raton Historical Society.*

Barrett, answered the door. She immediately came face-to-face with two military policemen (MPs). They told her that Mrs. Gladys Dixon, their neighbor to the north, reported seeing blinking lights coming from their direction. Insisting that the family's blackout curtains were in place and the household asleep, Jessie Barrett suggested they look at Dr. William Sanborn's empty home next door. The MPs roared away in their Jeep, followed by several heavily armed backups on motorcycles.

According to Peter Barrett, as the MPs approached the Sanborn home, they observed the back door ajar. Entering, they found wet towels in the bathroom, beds slept in, and in the kitchen, canned foods had been opened and consumed. "Pointing out to sea in the front of the house was a telescope and signaling light apparatus," Barrett said.

Discovering spies next door wasn't the only experience young Barrett had with the Germans. On another occasion, he became friends with one of the coast guard men who patrolled the beaches of Boca Raton every night on horseback. Learning of the man's upcoming birthday, Barrett asked his mother to bake a cake, and the next night, he took it with him to meet his friend. Unfortunately, the two never met. "I was told he was sick and couldn't come," Barrett said. "Later, we heard that there had been an encounter between the patrol and a German landing party on the night before his birthday and that there had been a shooting. I never saw my friend again, and we could never find out anything more about him."

Hitler's U-boat carnage subsided along U.S. borders in July 1942, after the initiation of a number of offensive strategies. The military quickly supplied planes patrolling the skies with new microwave radar that was undetectable by the enemy. With precision, the pilots picked off the U-boats. In addition, the navy policed convoys and worked in cooperation with military patrols, civilian observers, navy bombers, coast guard planes, blimps and crash boats—producing an elaborate communications and defense system. It proved an efficient combination that protected the Atlantic seaboard. This forced enemy submarines to move farther south into the Caribbean and Gulf of Mexico, where the U-boats sank an average of ten ships a week.

In her article "How Boca Helped Win the War," Waldeck noted, "Hitler and Admiral Karl Doenitz acknowledged that microwave radar was the principal factor in defeat of the U-boats."

As a conclusion to Operation Drumbeat, Wylie and Schwab wrote the following:

> *The war has been swept from America's doorstep. But we Americans should not forget that it was once there—close, menacing and real. Had England fallen in the air blitz, had Russia buckled at Stalingrad, the tide of blood and death that swept our own Eastern coast would surely have moved inland. Only when peace comes will the average seaboard-dwelling citizen realize the vulnerability of his community in the early months of the war and the debt he owes to what was actually a magnificent bluff—a bluff made by an armed services and carried out in part by civilians—but one that fooled the underwater professionals of Hitler's Reich.*

The Florida Division of Historical Resources Web site notes this interesting capture of German saboteurs:

> *Florida became the scene of a bizarre plot in June 1942 when four saboteurs came ashore from German submarine U-584 near Ponte Vedra Beach. They buried boxes of explosives and other equipment in the dunes for future use. The men then boarded a bus for Jacksonville, before splitting into two groups that traveled to New York and Chicago. The agents were to join with four other saboteurs, who had landed on New York's Long Island, and then planned to bomb key railroads, bridges and factories producing goods for the war. Fortunately, one of the New York band had misgivings about his mission and surrendered them to the FBI. By June 27 all of the men had been*

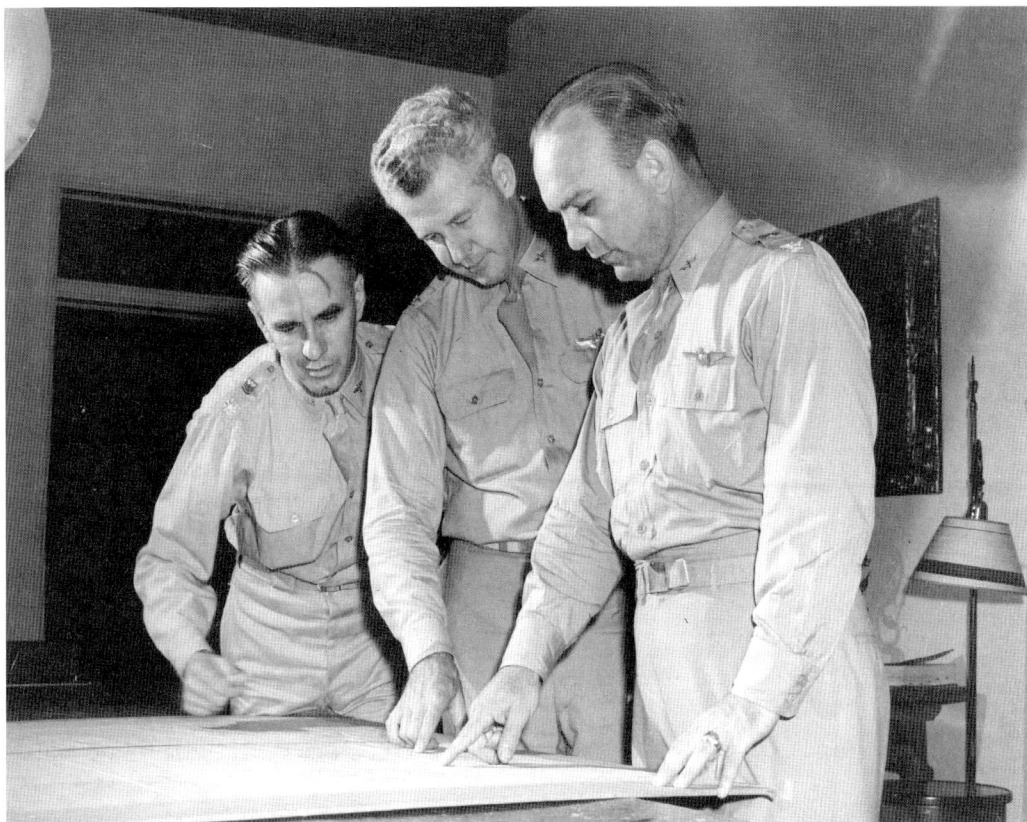

Colonel Thomas Bryan (right) reviews schematics of the new Boca Raton Army Air Field. He was instrumental in bringing the base to Boca Raton and later became its commander. *Photo courtesy of the Boca Raton Historical Society.*

> *apprehended. A military court later tried the eight Germans and found them guilty of spying. Six of the spies, including all of the Florida group, were executed.*

Is it possible these were the same brazen spies who invaded the Sanborn home just days earlier?

BOCA RATON GIVEN THE NOD

Aside from the obvious loss of life and disruption to commercial shipping, Hitler's U-boats wreaked havoc with Florida's tourist industry, one that relied heavily on the influx of northerners during the winter months. With lights dimmed or completely shut off by restaurants, nightclubs, hotels, apartments and shops, nightlife that once attracted tourists to the beaches was now almost nonexistent.

The horse and dog tracks soon closed, and stringent restrictions on gasoline, a result of the sinking of so many oil tankers, forced many tourists to stay home. Real estate agents began to panic for fear the season would be a disaster. Even the Boca Raton Club felt the pinch.

With German subs off Florida beaches and the military plunged headlong into the war, economic disaster loomed on the horizon. But Boca Raton Mayor J.C. Mitchell recognized the opportunity for his town to rise above the wartime doom and gloom. If it could be chosen as the site for the new Army Air Corps base, the economy of his town would reap substantial economic gains by catering to the needs of the military.

In a letter dated January 22, 1942, orders were issued at Scott Field, Illinois, for Colonel Thomas L. Bryan to transfer 27 officers and 352 enlisted men, together with their respective organizations, to temporary station at Morrison Field in West Palm Beach, Florida.

On February 5, 1942, Colonel Bryan—assisted by Captain Herbert C. Gee, Army Corps of Engineers, and Eli B. Hartner, medical corps—began site explorations in Florida. Wanting to keep the Signal Corps and Army Air Corps Technical School of Radar in close proximity, military officials considered three possible sites for the new airfield—Vero Beach, Fort Pierce and Boca Raton.

The *Boca Raton Radar School*, a series of six rolls of microfilm documenting the history of the Boca Raton Army Air Field, explains the result: "The endeavors of these three officers, supplemented by those of Majors Alvin G. Viney, Corps of Engineers, and James C. Moore, Medical Corps, took definite form on 2 March 1942 in an official recommendation to the War Department that some 5,820 acres of level, well-drained land located near the town of Boca Raton be acquired."

With this recommendation, negotiations at Vero Beach and Fort Pierce came to a halt: the former because of acquisition by U.S. naval authorities; the latter, for engineering reasons.

LAND ACQUISITION

On May 16, 1942 at the request of the secretary of war of the United States, U.S. District Judge John W. Holland in Miami, Florida, signed a petition of condemnation. This allowed the United States to acquire, by eminent domain, 5,820 acres located west of Boca Raton.

The land was bound by Dixie Highway on the east, the Seaboard Railroad on the west, Palmetto Park Road on the south and Fifty-first Street on the north. While no mention was immediately made of compensation to the landowners who would be displaced, the government conducted appraisals and paid property owners for their land.

Acquisition of this land was not an easy process. According to *Boca Raton Radar School* "A multitude of complexities delayed the final acquisition of the property. Over a hundred

Left: Hideo and Umeko Kobayashi in Japan in 1921 celebrating their arranged marriage. *Photo courtesy of Tom Kobayashi.*

Right: Tom Kobayashi in his WWII U.S. military uniform. *Photo courtesy of Tom Kobayashi.*

ownerships were represented within the area with fifty-one families in residence. A Negro community of forty families occupied houses and shanties that were built over a period of fifteen years on land not possessed through legal channels. Squatters were scattered in shacks. Settlement and/or compromise with numerous over-lapping taxing authorities was perplexingly involved."

Land was acquired from many families, some of who were immigrants from the Bahamas who had squatted on Yamato Colony land. Arthur A. Wells, one of these squatters, lived on this land. In an oral history commissioned by Expanding & Preserving Our Cultural Heritage Inc. (EPOCH) of Delray Beach, he said: "The Japanese lived at what is known now as Yamato Road. During that time we called it Jap[2] Town Road. There were Japanese families living out there and also black families who were employed by the Japanese. They didn't own the land. They were like what you would call today, squatters. All the land was government land. So they just moved in."

---

2. "Jap" was a common term and referred to persons of Japanese descent. At that time, it was not considered pejorative.

Nora Simmons was sixteen at the time she and her family were evicted from their home. Originally from Nassau in the Bahamas, her father, Alexander Knowles, and mother, Charlotte Knowles, relocated their family of thirteen to Yamato Colony land in 1928. There, the Knowleses picked vegetables for the Kamiya family, Japanese farmers of the former Yamato Colony.

"The government took the land and moved our homes to Delray Beach. It was nothing but woods when we went there," Simmons said.

The base displaced another squatter, Henry van Rolle. "Out of the clear blue sky, this man came there and said, 'Hello, you've got to move,' because our houses were close to the army camp. Well you know what they did? They paid for every house to be hauled away, and we all just moved."

The government purchased the homes and moved them to Delray Beach where the families formed a new neighborhood called New Town. The last of the original homes was razed in 2002.

Also in the path of the base was a cemetery, belonging to an association of plot owners. Located at NW Second Avenue and NW Sixteenth Street, the Army Corps of Engineers relocated approximately thirty-one caskets to an elevated mound south of Palmetto Park Road known as Sunset Hills. Today, the cemetery is called Boca Raton Cemetery and Mausoleum and is located at 451 SW Fourth Avenue.

A five-acre homesite and an additional fifteen acres owned by Hideo Kobayashi, one of the original Japanese settlers of the Yamato Colony, also occupied land requisitioned by the government.

Born near Kobe, Japan, Kobayashi came to Boca Raton in 1907 to join the Yamato Colony. In 1921, at age thirty-nine, he briefly returned to Japan for his arranged marriage to twenty-four-year-old Umeko. He brought her back to Boca Raton where they had four children.

After the colony disbanded in the late 1930s, Kobayashi bought several acres of farmland and raised beans, green peppers, eggplant, squash and tomatoes. Alongside daughter, Tomiko, and sons Sakaye (Theodore), Tomatsu (Tom) and Kiyoshi, Kobayashi picked vegetables and shipped them north in freight cars. The vegetables were also sold just to the south of Boca Raton at the Deerfield Beach and Pompano Beach Farmers Markets, both on the Florida East Coast railroad tracks.

Daily life for Tom Kobayashi was typical of most school-aged children. He was a member of the Boy Scouts, and his best friend was John Mitchell, son of Mayor J. C. Mitchell. He used to fish on Yamato Rock also called "Jap Rock," an outcropping on the beach just north of Spanish River Road. Fishing was good there, and he caught so many, he brought them back to the Montgomery Grocery Store where he exchanged them for food.

In 1937, Hideo Kobayashi gave up farming and became a landscape gardener. He started his own landscaping business, Kobayashi Landscape Inc. in Fort Lauderdale, and traveled U.S. Route 1 every day to go to work. Then the government's eviction notice arrived.

Tom, who was fourteen at the time, gives his account of that day: "The FBI came by in the morning with a warrant to search the house for anything they wanted. They took a shotgun, radio, books on Japan and a small toy signal set, a present when I was younger that I had thrown into the attic. They told me, 'Did you know you could send messages to the German subs a couple miles away with this?' They confiscated it."

The Kobayashi family wasn't placed in an interment camp like Japanese on the West Coast. Instead the military relocated their home, built of wood frame with shingle siding, to Spanish Village just north of the town hall. "They only gave you what they thought it was worth. A lot of people who had their property taken were mad because they weren't given the proper value," said Tom Kobayashi.

The coast guard dispatched two men to the house to watch the family. Tom recalls one of the guards, a pure-blooded German, asking him, "Why am I guarding your dad when nobody's guarding my dad in New York?"

The military restricted Hideo Kobayashi's movement. In order to travel, including to work, he had to obtain a permit from the courthouse in West Palm Beach. Even then, a guard accompanied him. Shortly after their relocation to Spanish Village the Kobayashi family moved to Delray Beach where they rented a home. But Hideo found the commute to work too far in his 1936 two-door Ford sedan, so after a short six-month stay, he moved his family to Fort Lauderdale. Once there, the military removed coast guard protection.

"We never saw the base, and never went back after we left," said Tom Kobayashi.

The Kobayashi family wasn't the only one to feel the pain of displacement and sting of inadequate compensation for land and home. Drollene Brown, in her article "World War II in Boca Raton: The Home Front," wrote: "Another piece of property taken over by the military was the building west of the FEC [Florida East Coast] railroad tracks where the Louis A. Zimmerman family had lived and had a restaurant before starting ZIM's on Federal. 'My father was paid $2,500 for the two-story nine-room building,' said Lucille [Zimmerman Morris]. 'After the war, he was given the chance to buy it back for $6,000, but he refused.'"

BOCA RATON CLUB COMMISSIONED FOR WAR EFFORT

Along with the acquisition of land for the base, the government issued an order taking possession of the exclusive Boca Raton Club. While the base was under construction, the club became the temporary home for thousands of troops, officers and cadets (officer trainees) who poured into her hallowed hallways, ornate dining rooms and guest suites.

Surrounded by two golf courses, six tennis courts, swimming pools, beaches, cabanas and grounds world-famous for their tropical horticulture, the Boca Raton Club followed Scott Field and Morrison Field to become the third home of the Army Air Corps Technical School of Radar. Colonel John W. Monahan assumed duties as the first acting commandant.

A golf course with lush tropical foliage surrounded the private Boca Raton Club. *Photo courtesy of the Boca Raton Historical Society.*

According to the *1947 Army Air Forces Training Command Year Book*, donated to the Boca Raton Historical Society archives by Captain A.J. Mills: "The school was moved from Morrison Field to the aristocratic Boca Raton Club, where amid the unmilitary atmosphere of beach cabanas, golf courses, and tiled luxury, the school functioned until the new air field was completed."

In the midst of this transition, acting club manager, Harold Turner, directed staff to store furniture and place coverings over the carved plaster pillars for protection. While there, the troops used the pool, beach and dining facilities. On the golf course, they practiced maneuvers and dug foxholes.

In January 1943, Wilbur Bell, along with five fellow aviation cadets, boarded a train in Portland, Maine, and three days later arrived in Boca Raton. Getting off the train at what he described as a "small barn with no one in sight," they sat on the curb until an army truck arrived to ferry them to the Boca Raton Club.

"Our cadet group was the second group to be sent to the club. Our first duty was to remove all the rugs and clean the cement floor with GI soap and razor blades to make the floor shine. It shined," said Bell.

Across from the Boca Raton Club, the Cabana Club was directly on the beach. *Photo courtesy of the Boca Raton Historical Society.*

He recalled that the cadets at the club were under the impression their current quarters had previously been used "as an exclusive club for the Chicago mafia"—although he didn't know how that rumor got started.

That rumor may have been true, given a 1942 article by Judith Cass (publication unknown), in which she wrote: "Those wealthy Chicagoans who are members of the fabulous Boca Raton Club in Florida will not have a chance to spend vacations there next winter, for the entire east coast resort, including the Club, has been taken over by the Army."

While living conditions were sparse, officers did have private rooms. The beds were the "painted wood headboards used by the hotel, but there were no rugs in the halls, carpets in the rooms or pictures on the walls." said Manuel "Manny" Chavez, an officer stationed at the club. Today, one of the original headboards is on display in the library of the Boca Raton Historical Society.

In the back of the club was a temporary post office manned by the military, but by October, a new postal building opened and operations were transferred to the on-base facility. On October 26, the field post office became an independent branch of the Fort

Cadets receive instruction on the golf course of the Boca Raton Club. *Photo courtesy of the Boca Raton Historical Society.*

A bare interior—devoid of furniture, carpet and artwork—greeted new cadets at the Boca Raton Club. *Photo courtesy of the Boca Raton Historical Society.*

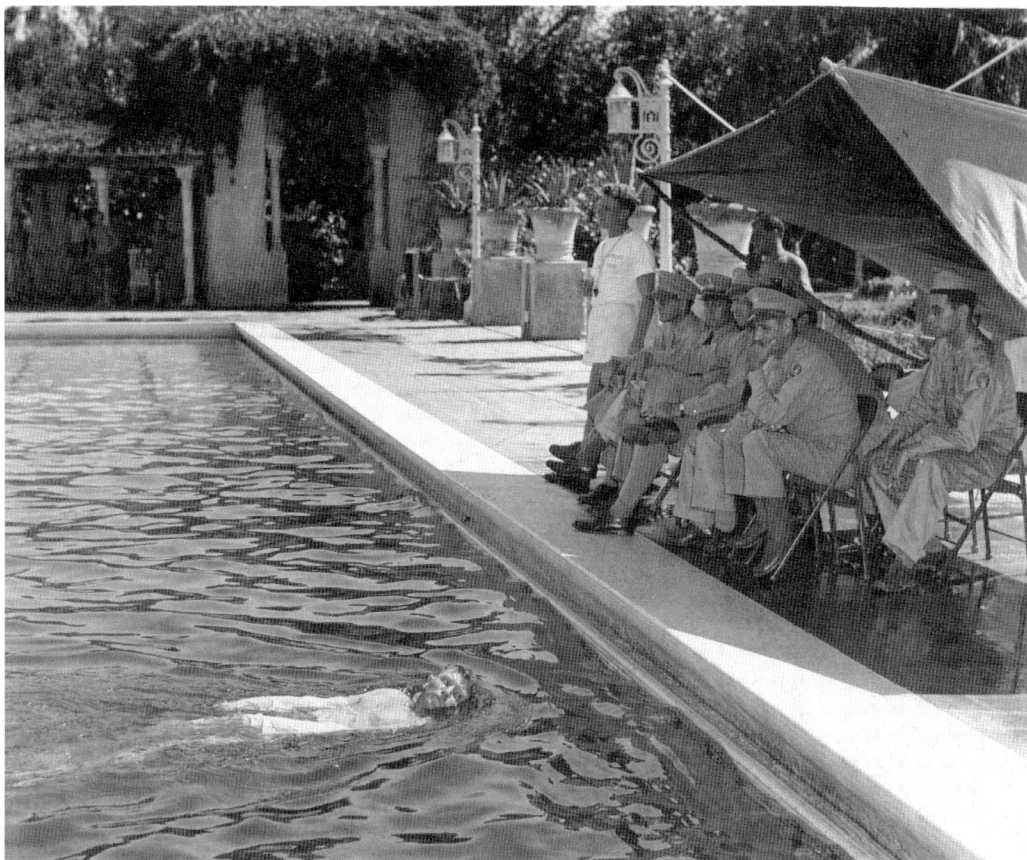

Troops used the club pool for recreation and training. *Photo courtesy of the Boca Raton Historical Society.*

The *Transmitter*, the official base newspaper, put out its first edition in July 1942. *Photo courtesy of the Boca Raton Historical Society.*

Lauderdale post office, handling an average of 400 packages daily. By December 15, daily volume of parcel post reached 3,200 packages.

In July 1942, a base newspaper was authorized and seventeen days later personnel read volume 1, number 1 of the *Transmitter*. The weekly paper was distributed to members of the command without charge.

The Special Services Office acquired all the content for the paper and the *Delray Beach News* financed the publication through the selling of space. Mr. Lauren Hand published the first twelve issues as a civilian enterprise, but the paper gradually expanded to eight pages and publication transferred to Fort Lauderdale.

At the rear of the Boca Raton Club, the military built a temporary stockade (military jail) that consisted of four pyramid-shaped tents mounted on floored frames. An eight-foot wire fence, garnished on top with four strands of barbed wire, encircled the tents. Later, the stockade relocated to the Boca Raton Army Air Field.

Gerald Weiss, who was put on stockade duty at the base, had this to say about his guard time: "I was assaulted by all kinds of huge flying somethings that acted as dive bombers while in the elevated guard towers." Locals knew these as Palmetto bugs, a several-inch-long American cockroach.

Enola Gay Countermeasure Officer Stationed at Boca Raton Club

Born in Baltimore, Maryland, Jacob Beser studied engineering at Johns Hopkins University. The day after Pearl Harbor, he enlisted in the air corps and attended the Army Air Forces Technical Training Center at the Boca Raton Club from October 5, 1942 to January 17, 1943. He graduated from radar school as a lieutenant and was initially assigned to Orlando, Florida.

As radar-countermeasure officer, Lieutenant Beser was the only man who flew aboard both strike planes that dropped atomic bombs on Japan, ending World War II. Handpicked by pilot, Colonel Paul W. Tibbets, he flew aboard the *Enola Gay* on its historic run over Nagasaki.

As the aircraft's radar-countermeasure officer, Beser's responsibility was to ensure there weren't any electronic signals in the target areas that might interfere with the proper operation of the electronic proximity fuses on the weapons. He was also trained to detect any emitted surface-to-air signals that might prematurely operate the fuses and detonate the bomb, destroying the crew and aircraft.

For his service, Beser was awarded the Silver Star for "distinguished gallantry in action against an enemy of the United States or while serving with friendly forces against an opposing enemy" as well as the Distinguished Flying Cross for "persons who while serving in any capacity with the Armed Forces of the United States distinguishes himself by heroism or extraordinary achievement while participating in aerial flight."

"It seems quite probable that Sgt. Joseph S. Stiborik [*Enola Gay* radar operator] also attended school there [at Boca Raton Army Air Field] but we cannot locate confirmation

Lieutenant Jacob Beser poses in front of the *Enola Gay* at Roswell, New Mexico. The symbols on the plane are "Fat Men" and indicate missions the plane flew. *Photo courtesy of Leon Smith.*

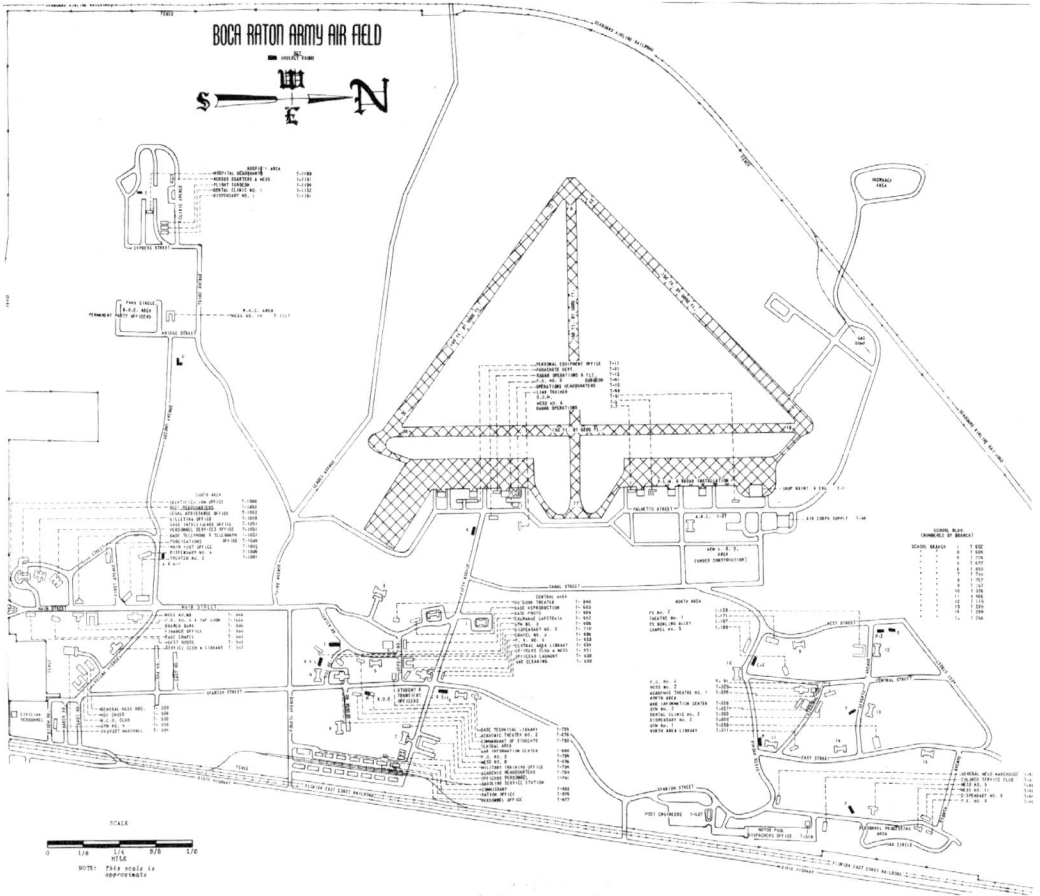

Although all structures are not shown, this schematic depicts the initial layout of the base. *Photo courtesy of the Boca Raton Historical Society.*

of this in our records. Both members [Beser and Stiborik] are now deceased, as is Pfc. Richard H. Nelson, the radio operator. So far as we are aware, only three members of the crew survive," wrote Dan Hagedorn, archivist and adjunct curator, Latin American Aviation, Smithsonian Institution, National Air and Space Museum.

THE BASE TAKES SHAPE

With the Boca Raton Club housing the Army Air Corps Technical School of Radar and the acquisition of base land completed, construction of the airfield commenced and a name for the base was sought. Post Commander Colonel John W. Monahan recommended that the field be named Griffiss Field in honor of the late Colonel Townsend Griffiss who lost his life in an airplane accident while abroad. Although the army rejected this

Ed Wallish poses beside his Student Squadron T0-2 sign. *Photo courtesy of Ed Wallish.*

name, a year later correspondence was still being addressed to Griffiss Field, as Monahan had used the name on official correspondence before it was rejected.

On June 2, 1942, the Army Air Forces Naming Board gave the base the temporary name, Boca Raton Field. However, in October, several cities, towns and civic organizations in the vicinity petitioned then-Post Commander Colonel Lawrence J. Carr to rename the field in honor of the late Second Lieutenant Alexander H. Nininger Jr. A member of the Fifty-seventh Infantry, he had lived in southeastern Florida and died in heroic action against the enemy near Abucay, Bataan, in the Philippine Islands.

Once again, the naming board rejected this suggestion, citing paragraph 9, section II, Army Air Force Regulation No. 200–1, which reads: "The name of any individual recommended will be that of a deceased flying officer of the Army of the United States." Thus, the name, Boca Raton Army Air Field, took hold and later became official.

Heading up the construction crew for the Boca Raton Army Air Field was veteran military engineer Colonel Arnold MacSpadden, who was under the impression the base was to be a "temporary twin post to the Army Signal Corps Station at Camp Hobe Sound," located seventy miles to the north.

Bill Robison, who served at the Boca Raton Army Air Field in 1945, explained the difference between the Signal Corps and the Boca Raton field as well as how their missions complemented each other:

> There were numerous Army Air Force Bases throughout the United States whose mission was the training of Army Air Force (AAF) personnel in the various technologies required to operate and maintain the equipment required to support the AAF during WWII. To the best of my knowledge, BRAAF [the Boca Raton Army Air Field] was the only military facility that was created for, and whose sole mission was, the training of AAF radar operations and technicians. It should be noted that the training was not only in radar. It was directed to the operation and maintenance of virtually all of the airborne electronics equipment being utilized by the AAF.
>
> Whereas BRAAF was the only facility dedicated to the training of AAF radar operators and mechanics, there were other military radar training facilities. The most notable of which was Fort Monmouth, New Jersey, which was an Army Signal Corps facility. Fort Monmouth's mission was the training of operators and mechanics who were to operate and maintain all electronic equipment that Army Ground Forces utilized in support of their mission. This included radar for overhead aircraft detection and location and the direction of anti-aircraft gunfire.
>
> Fort Monmouth existed long before the onset of WWII, and thus was regarded as a much more high-tech electronic facility than was BRAAF. Thus Fort Monmouth was involved in the early development of much of the radar that BRAAF ultimately was assigned. This equipment was used to train the vast number of operators and mechanics projected to support the air war in Europe and the Pacific. There were, in fact, a significant number of BRAAF trainees that were sent to Fort Monmouth for specialized training.

Command headquarters fast-tracked the building of the base, and in four short months, eight hundred buildings emerged out of what was once white sand and palmettos. Construction of the runways and buildings utilized 14 contractors, 12 of them from Florida, and 3,500 construction workers from nearby cities such as Fort Lauderdale, Delray Beach and West Palm Beach. Construction costs totaled $12 million.

"The primary mission was to train airborne radar operators and radar mechanics, but schools were also established in high and low altitude bombing, radar navigation, airplane identification, attack interception, and survival training," Dick Kelley wrote in the June 19, 1961 *Fort Lauderdale News* article "Joe College To Take Place of GI Joe."

Veteran Ed Wallish gives a physical description of the base: "The main base was divided into three areas: the north, central and south. The airfield was located to the west of the central and north areas, the base hospital was located on the southwest part of the base and the town of Boca Raton and the Atlantic Ocean were to the east."

A 1942 aerial view of the Boca Raton Army Air Field. *Photo courtesy of the Boca Raton Historical Society.*

Paul Metro stands in a field of sand and palmettos outside the barracks. The foliage was left intact to help camouflage the base. *Photo courtesy of Paul Metro.*

"Most of the other buildings on the base were of wood frame, tar-paper-covered construction. There were movie theaters, mess halls, airplane hangars, gymnasiums, post exchanges [similar to a department store], fire stations, office buildings and the hospital," Wallish continued. "There were some warehouses of concrete construction and L-shaped, cinder-block barracks near the flight line. The H-shaped [radar] school buildings were also constructed of cinder blocks."

Consisting of four runways, the triangular-shaped bituminous airfield was the centerpiece of the base. One 6,400-foot-by-150-foot strip ran north to south; three 5,000-foot-by-150-foot strips ran northeast to southwest, east to west and northwest to southeast. The remainder of the base was constructed east and south of the airfield. Trees and shrubbery, left intact, helped camouflage the facility. Hard lime rock and lime slurry, found at Deerfield Beach, became the foundation for twenty-five miles of roads laid out in a winding pattern rather than the typical crosshair grid of most military bases.

Some accounts of the construction period note that the winding roads and haphazard placement of the buildings were because of construction deadlines; it took too much time to remove the obstacles so the engineers built around them. Other accounts note this was done so that the Germans couldn't blow out the whole system at once should the base come under air attack.

Kenneth Jacobson, who was stationed at the base during its construction phase, stated in a May 29, 1983 interview in the *Fort Lauderdale News/Sun-Sentinel*, "The idea was to build two of everything—two water systems, two electrical systems—so that if one got blown out, we still had the other."

The water supply system was divided into two complete and interconnected units, one located in the north area and the other in the central area. Six wells were sunk two hundred feet into the limestone. Engineers added chlorine at the pumping stations and hardened the water with chemicals.

Florida Power and Light supplied electric power. The sewage collection system consisted of nine gravity sewer networks. The complete network consisted of 19.9 miles of piping, nine lift stations and 281 brick manholes.

Buildings were first constructed of pine, but Colonel MacSpadden later resorted to using concrete blocks that he found in Miami after the pine lumber proved to be the wrong type and size. Each structure was designated by the prefix "T" for temporary, followed by a number that corresponded to the legend on official construction plans (e.g., T-3 and T-615). When troops arrived, the buildings were typically referred to by their use, such as the chapel, BOQ (Bachelor Officer's Quarters) or day room, but others were referred to by their T number.

Plans called for construction of three chapels, and on October 4 construction crews completed the first one. By year-end, with all three completed, Hammond organs were installed. The chapels measured thirty-nine feet by ninety-seven feet wide and thirty-five feet high. They seated 280 on the main floor and 50 in the balcony. Two small

Three chapels, complete with Hammond organs, were built on the base. *Photo courtesy of the Boca Raton Historical Society.*

A couple weds in a base chapel. *Photo courtesy of the Boca Raton Historical Society.*

offices occupied either side of the front entrance, and two bathrooms were located at the front and rear of the chapel.

The structures—of frame construction with high, arched wooden beams—were purposely designed to be a place of beauty where, according to the *Boca Raton Radar School*, men on the field could find "a moment's relaxation at any time of the day." Services were held on Sundays for the various religions, and periodic weddings and baptisms performed.

Eight branch retail outlets, commonly known as the PX (or "post exchange"), were in operation by the end of 1943. They were located in the north, south, Section F (the colored section), hangar line, hospital and one each in the two embarkation areas. The larger branches contained a civilian-operated laundry and tailoring service.

Meals were served out of eleven mess halls located in various sections of the field. The mess halls were so far apart, to drop off supplies to each required traveling a distance of 13.5 miles Seating capacity was 7,625, and troops ate in shifts. A separate mess hall was maintained in the nurses' quarters for the Army Nurse Corps.

Before construction of the base hospital, troops at the Boca Raton Club depended upon medical facilities at Morrison Field in West Palm Beach, twenty-five miles to the north. However, on July 6, 1942, a small medical contingency arrived on the base, and by the end of December, 51 officers, 8 nurses and 330 enlisted men served 8,708 military personnel. The hospital had all the necessary medical equipment and supplies for general medical and surgical care. Six dispensaries (smaller clinics) were established and opened in various sections of the field.

With most of the structures in place, on October 15, 1942, Post Commander Colonel Carr formally opened the Boca Raton Army Air Field. However, his command was short-lived. On November 13, 1942, Colonel Thomas L. Bryan, the officer instrumental in development of the Boca Raton Army Air Field and the technical training school, assumed duties as commander.

TROOPS SENT TO BOCA RATON ARMY AIR FIELD

During 1945, when Bill Robison trained at the Boca Raton air field as a radar mechanic, he recounted that "virtually all of the men were 'washed out' cadets," including himself. According to Robison, these men originally enlisted in the Army Air Force to be trained as commissioned officer air crewmen—pilots, navigators and bombardiers. He explained:

> *During 1942 and 1943 when the AAF was experiencing staggering losses of aircraft and crews while carrying out the daylight bombing raids over Europe, recruiting teams visited high schools throughout the United States to administer written tests directed [toward] selecting and signing up young men of seventeen to join the AAF enlisted reserve. The understanding was that upon turning eighteen, they would be called to active duty, and after basic training, would be trained in one of the aircrew*

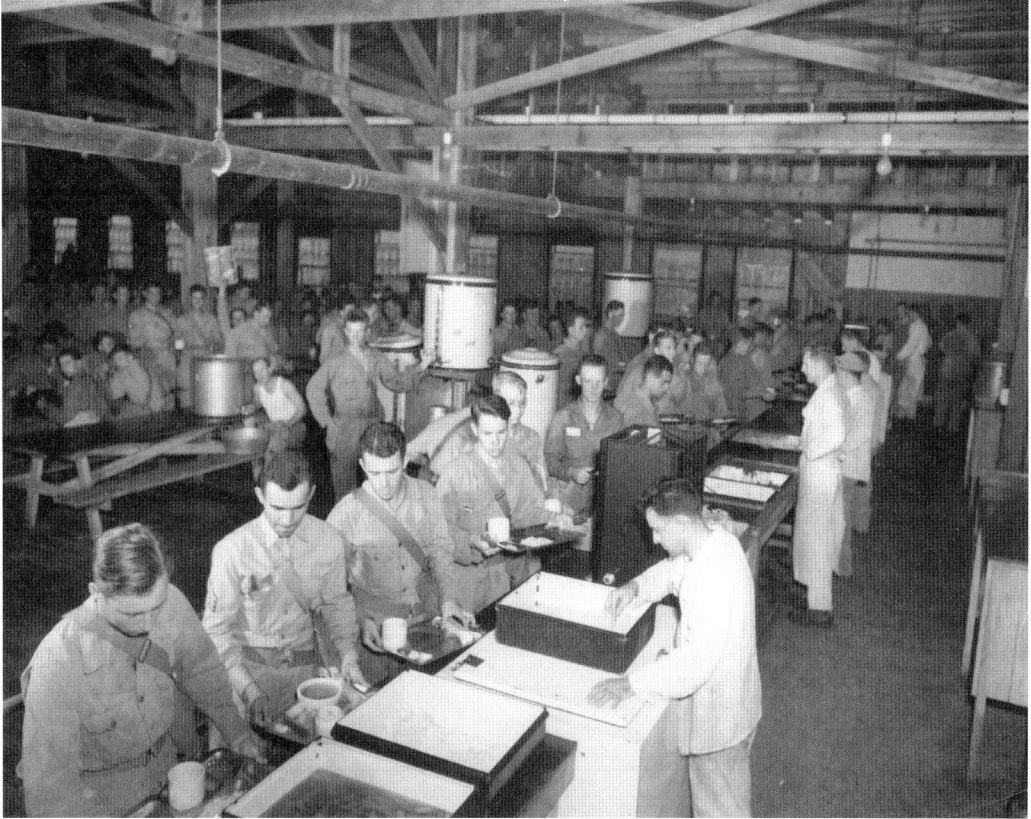

Troops file through one of the Boca Raton Army Air Field's eleven mess halls. *Photo courtesy of the Boca Raton Historical Society.*

specialties. Upon successful completion of that training they would be commissioned as second lieutenants and assigned to combat squadrons.

As it turned out, by 1943 the tide of the war was beginning to shift, and the Allies were gaining superiority, both on the ground and in the air. In fact, by 1944 our losses of aircraft and aircrew had dropped dramatically, and thus the projected need for replacement crews dropped dramatically. As a consequence, the pre-flight screening and training of the air crew cadets was made even more rigorous, and vast numbers of cadets were "washed out" of flight training, many before they came anywhere near a training aircraft. I was among this group.

There continued to be, however, the need for enlisted air crewmen. These were the aerial gunners, radio operators, radar operators and aircraft engineers who were required aboard the B-17s, B-24s and later, the B-26s, that were still flying missions over Europe and the Pacific. Thus, the "washed out" cadets were reclassified and sent to AAF bases all over the country to be trained in the various specialties as enlisted air crewmen.

It is possible the barracks were intentionally laid out in a helter-skelter pattern to thwart potential attacks from the Germans. *Photo courtesy of the Boca Raton Historical Society.*

According to Robison, it was understood among the troops that those "washed out" of radar training bore the distinct possibility of being trained as an aerial gunner, a position with a high mortality rate. Before knowing where the "washed out" cadets were to be assigned, they composed a song to the tune "Stars and Stripes Forever."

> *Be kind to your washed out cadet,*
> *For he may be your aerial gunner,*
> *He lives in the waist and the tail,*
> *Where the bullets fly like hail,*
> *Now you may think that he will be killed,*
> *Well, he will.*

Troops arrived in Boca Raton, mostly by train, from every corner of the country. Those traveling from the West changed trains in Jacksonville, Florida, and boarded

Left: Bob Davey poses in front of his tarpaper barrack. *Photo courtesy of Bob Davey.*
Right: William Eddinger served from 1942 to 1946 first as a trainee, then as an instructor. *Photo courtesy of Pat Eddinger Jakubek.*

Troops washed their clothes on large, flat scrub tables outside the barracks. *Photo courtesy of Bob Davey.*

the Seaboard Airline that headed south along what is now the Interstate 95 corridor. Troops from the East boarded the Florida East Coast railroad (Henry M. Flagler's FEC) that parallels Dixie Highway.

Train accommodations varied among the troops. Some traveled by Pullman Coach, a posh method of travel for a soldier, while others experienced the close quarters and decidedly understated accommodations of a troop train.

Ned Renner arrived in Boca Raton via troop train, "happy to get off of that train after four days and nights of no bathing facilities and having to eat out of mess kits."

John H. Cochrane shipped out to Boca Raton from electronics school at Chanute Field, Illinois, in February 1945. While he knew where he was going, he recalled very distinctly being told that "no one, not even parents, were to know where we were going and why."

He was awed by his arrival in Boca Raton and explained: "Imagine just having left the subzero temps of a Northern winter and waking up one morning just as we ground to a screeching halt at the Boca AAF rail siding adjoining Dixie Hwy. I lifted the Pullman's window shade to see—PALM TREES! Greenery! Sunshine! This Rock Island rube, who had never been farther south than Illinois, was most impressed. And the white sand and balmy air! This was war? I resigned myself to make the best of it."

Military personnel disembarked at the Florida East Coast Seaboard depot in Boca Raton, the Seaboard depot west of the base or at depots in Delray Beach or Pompano Beach. Many times, troops were met by base personnel, but some arrived to empty platforms. After making their way to the base and checking in with their commanding officer, the men were issued bedding, accompanied by a complimentary mosquito net, and assigned a barrack.

Paul Metro arrived from Truax Field Mechanic School. He described his new living quarters as "one-story, wood barracks laid out helter-skelter, not in the typical military style of rows. They were built on concrete piers with iron anchors as protection against hurricanes. ... The roofing was of two or three colors of camouflage. The ground was sandy, covered with scrub pine and palmetto bushes, with boardwalks leading from the barracks to the streets."

Bob Davey said the layout "was as though you picked up a handful of dominoes and threw them on the floor; whichever direction they were facing that's the way the barracks were. There were all different angles, spread different distances apart, supposedly bomb proof and attack proof."

According to Renner, the interior of the structures was "an open space you could see through from one end to the other with the only enclosed area[s] in any barrack being the latrine and the other a small room for the non-commissioned officer in charge of the barrack."

Outside the barracks stood large, flat scrub tables made of wood where the men washed their personal items. Bed linens and uniforms were sent out to local laundry facilities.

Without air-conditioning, Harry Magafan recalled, "the barracks were stifling in the summer." To combat the loss of salt from perspiring troops, buckets of salt tablets were left beside the water fountains.

One of the early 1943 arrivals, Joe Ornstein, claimed he didn't live in the luxury of a barrack but in the woods in a Nissen hut. He described the hut as a "sort of a round metal place" and the surroundings as "wild." Said Ornstein, "Outside you would see alligators and things like that and you had to watch how you got into the hut."

Sergeant William R. Eddinger was one of the first to train at the Army Air Corps Technical School of Radar in 1942 at the Boca Raton Club. His daughter, Pat Eddinger Jakubek, a Florida resident since 1943, gives this account of her father's impression upon arriving for duty at the newly built base: "When my father arrived at the Army Air Corp base atop the hill just west of 'downtown,' he saw that this was a different land. Having lived his first twenty-five years among central Pennsylvania's rolling green hills, sparkling rivers and creeks, fertile farmland and conservative demeanor, my father was amazed at all he saw.

"What he saw was flat land, which ran as far as the eye could see, covered with white sand and scrub pine," Eddinger Jakubek continued. "Coconut palm trees, heavy with fruit tucked beneath the fronds, swayed to even a whisper of a breeze. Waves of white heat danced on blacktop pavements. Sandspurs were everywhere ... in the brush, at the beach, in the little patches of green surrounding the homes that made up Boca. And he saw the beach, the wonderful beach, the golden sand bowl cupping the salty ocean. I believe he'd seen a good sized lake or two in his time, perhaps even one of the Great Lakes, but it must have taken his breath away when he saw the wide expanse of the Atlantic Ocean."

DISCRIMINATION AND SECTION F

During the 1940s, segregation touched every facet of American life. The military was no exception. In the North, blacks and whites grew up together. In the South, Jim Crow segregation was an accepted way of life.

William Phears, who became one of the Tuskegee Airmen, wrote about his experience with segregation in 1943 as a new cadet at the Boca Raton Club. In his book, *Ain't, But It Can Be*, he wrote:

> *Each black cadet had to be concerned about his performance, the protection of his pride, adjust to the social discomforts, face racial encounters and use strategic and tactical schemes to enjoy all the things enjoyed by others. It is without exception that each of these trainees experienced the raucousness of the formal policies of segregation. It was self-confidence, and determination that motivated all of the selected blacks to endure the treatment, fulfilling the training requirements, and perform effectively. The black cadets had a very closed network. Communication was maintained with new individuals who were known to be exceptionally qualified to help each other avoid pitfalls and exclusionary tactics.*
>
> *Occasionally racism provided a release for the oppressed. During the indoctrination of a new class of recruits the policy was emphasized that every new cadet was restricted*

Above and below: Going about their daily routines, Squadron F, the colored squadron, was segregated from the rest of the base. *Photo courtesy of the Boca Raton Historical Society.*

to the base for the first four weeks and that every cadet had to get a haircut every week. After four days, three of us (new black cadets) proceeded to the cadet-center barbershop. All the barbers (who were white) stopped work, the military police were called and we were invited into the commander's office. The commander asked, "Just what are you cadets trying to do?" We explained the policies and asked if we did not have to comply. He then asked: "Where are you from?" (pointing to each cadet) The responses were: "Chicago" (Williams), "Detroit" (Phears), "New Orleans" (Brown). The commander shouted: "You should know better!" We just sat and watched him as he called the Adjutant and stated that we could be excused every Saturday, after retreat, to go to Palm Beach or Ft. Lauderdale to get haircuts. Of course we enjoyed a little social life after leaving the barbershop.

    Curfew on Saturday night was midnight, but because of limited available transportation, we were allowed to break curfew. Parenthetically, before training

Archie Carswell (right) and a friend served in Squadron F.
Photo courtesy of Archie Carswell.

was curtailed at Boca Raton, the incoming black cadets were able to get their hair cuts in the hotel.

In addition to the above experiences, Phears and the other black cadets never rode the same bus or sat together. Neither did they ever play on the same sports team. At a squadron dinner party in Fort Lauderdale, hotel management told the colored squadron to sit at a separate table near the kitchen. Embarrassed by this treatment, the commandant, chaplain and several cadet officers, all white, joined their black colleagues at the table by the kitchen door.

Phears wrote: "The local civilian power structure began expressing concern about the black cadets seeking full equality when off the military installation. After about one year, this Basic Training School was moved to Seymore-Johnson Air Field, North Carolina in April 1944, although the physical facilities and climate were perfect for training at Boca Raton, Florida."

On the Boca Raton Army Air Field, coloreds were assigned to Squadron F and housed in Section F, located in the northeast area of the base. The members of Squadron F had their own PX and on-base entertainment. While the 503rd AAF base band was integrated, few other venues allowed blacks to mingle with other personnel, military or civilian. Off base, blacks frequented segregated bars and entertainment spots.

Archie Carswell served on the base from 1942 to 1943. As an enlisted man (EM), his job was to transport new equipment from its arrival point off base to the processing point on base, and drive cadets to and from the train depot and Boca Raton Club. The mode of transportation consisted of long, open trailers that were towed by trucks. Benches ran down both sides and the middle.

Of his experience as a soldier in Section F, Carswell said:

I had to take the truck I usually drove soldiers in and use a winch to pull down a water tower that used to be on the Japanese man's land [Hideo Kobayashi]. It was a beautifully carved thing, and I hated to do it. They kept the Japanese separated like they kept the black soldiers separated from the whites. They did not trust the Japanese, so segregation made me feel the whites did not trust the blacks.

We were all on the base there together, but we could not eat at the same place and we could not have recreation or training together. It made us wonder how you were going to fight a war and expect a man to put his life on the line for America. I felt that we were both fighting the war so we should fight together.

It is different now, but I am sure there is still prejudice in the army. To try to get a stripe, you'd be tough on your buddy to get a promotion real quick then they'd know you wouldn't take anything off of anyone. We called it "eat some cheese," that meant "cheese" on your friends, your buddy.

We just accepted what we had to accept, that was about all we could do. There was a world war, you just wanted to do what you had to do and get out.

Ned Renner recalls his first introduction to segregation:

The U.S. Army was segregated at that time, and I had no contact with black soldiers. Nor did I have any knowledge of segregation in the South United States. The first time I ever boarded a public transportation bus in Florida, and I am sure that was in Fort Lauderdale, I walked to the back of the bus and sat down. There were a couple of blacks sitting in seats in the back also. After some long moments, I wondered why the bus was not leaving and noted the driver looking back at me. He finally came back to me and told me I would have to move to the front of the bus because I was white. That was my first real lesson in racism. I did not understand it, but I complied with the driver's request. I did witness a couple of race riots while I was in basic training and really did not know what that was all about.

Ed Wallish said he never saw black GIs in radar training but saw plenty in the mess halls pulling kitchen patrol (KP) duty. He explained:

At the time, soldiers were kept segregated by race, and at Boca, all the KP was being done by black soldiers. It was rumored that these soldiers were originally sent for radar training. But, because they were black, they weren't allowed to start school and were waiting to be shipped out. While awaiting shipment, they were all casuals assigned to doing the KP for the base and were extremely unhappy about it. I believe there is much merit to the rumors as there weren't any black radar students at Boca while I was there. Eventually all the black soldiers were shipped out. The KP then was taken over by casuals in the student squadrons.

Ken Davey (far right) poses with his flight crew before a bombing run. *Photo courtesy of Colonel Kenneth W. Davey USAF (Ret.).*

RADAR AND THE WAR IN EUROPE

With the Army Air Corps Technical School of Radar now open, radar training was underway. Up to this time, relations between the Rad Lab in Massachusetts and the military were strained, at best. The military wasn't fond of being excluded from decisions that affected troop training and operations. By the summer of 1942, the CXBK, an experimental microwave ASV (air-to-surface vessel) radar, was in operation and its scope clearly defined coastlines and bays, towns and cities. Because of its precision, Royal Air Force dignitaries declared it "a turning point in the war."

To put radar training at the Boca Raton Army Air Field and its use in the field in proper perspective, Colonel Kenneth W. Davey USAF (Ret.)—a B-17 pilot stationed in Molesworth, England, with the 303rd Bomb Group, and in Polebrook, England, with the 351st Bomb Group (1942–1944)—gave a firsthand account of events leading up to the use of radar in military aircraft flown by the Eighth Air Force, where many of the men from the Boca Raton air field were assigned after their training.

In September of 1942 the US Army Air Corps became involved with the air war in Europe by sending four B-17 bomb groups to England to help the Royal Air Force [or RAF] in the war with Germany. The RAF had been bombing Germany for some time but their tactics only allowed for area bombing by dispatching bombers at night. Each bomber was dispatched individually and flew in trail [behind one another] *to the target area. They had to fly at night because they weren't heavily armed and had to use darkness to protect themselves from the Luftwaffe defense fighter forces.*

The U.S. Army Air Corp (AAC) had a different approach to the bombing program—build a big bomber aircraft, equip it with a bombsight that could be used in day light and drop bombs on specific small targets with great accuracy. But if the operation needed daylight to succeed, the aircraft would be exposed to German fighters. To reduce vulnerability to fighter attacks, the AAC put 50 caliber machine guns on the aircraft for its own defense. To further reduce vulnerability, the aircraft flew in groups that would provide better protection by bringing more 50 cal. guns in range to fire on approaching enemy aircraft.

These four bomb groups entered the war in the fall of 1942, having to prove that it was feasible to bomb Germany by daylight—feasible meaning, measuring bombing effectiveness against crew attrition. After some experimenting, a tactic was accepted. Each bomb group would form a flying formation of 18 aircraft bomb groups who would follow in trail to provide enough bombs for target distribution. Groups assisted each other with a maximum amount of firepower used against intercepting German fighter aircraft.

After six months of testing, the system proved feasible and the 8th Air Force grew into a great air armada. While all this was going on, there was dissatisfaction with bombing accuracy and the cost effectiveness of the whole program. As protection from German anti-aircraft flack, planes had to bomb from high altitudes. The bombardiers also had to see their targets and hence couldn't drop their bombs if cloud cover interfered with their vision. Such failure meant flying a several hour mission and returning home with unused bombs or using them, probably ineffectively, on other targets of opportunity.

While time passed there was experimentation taking place back home between England and the U.S. They were sending electronic beams out into space to strike an object and reflect back to the sender. The sender had a screen that received the reflected beam that would appear as black dots. If the dot was moving, he could watch it on his screen and tell how fast and in what direction it was moving. The system took on the name of RADAR and equipment was built to use it to spot aircraft that might be approaching to drop a bomb or fire a round.

The advantage of radar over eyesight was radar could look through clouds and give the sender a picture of what was happening on the other side. This soon became

B-17s flying in formation on a bombing run. *Photo courtesy of Colonel Kenneth W. Davey USAF (Ret.).*

our secret tool for ground air defense. If we could use it to look up through clouds why couldn't we use it on bombing aircraft so the bombardier could look down through a cloud layer, see his target and drop his bombs? This was the beginning of airborne radar bombing.

In 1943 radar guns were mounted with the bombsight. Naturally the picture he saw was the radar beam reflected from the surface below and the density of the surface created the picture. Highways, rail roads, airport runways and factory roofs returned black pictures but trees, pastures, grasslands, lakes and rivers were seen as various hues of gray. The bombardier was now a member of a lead crew in his bomb group and he was sent to a "lead crew bomb school" run by the 8th Air Force.

The school had radar pictures (photos) of target areas and the bombardier, upon arriving at his initial point, took over the navigation of the aircraft to the target. He had to match what he saw with the radar photo to his screen and when they became "one" he dropped the bombs and the rest of his formation dropped on his signal.

With the success of airborne radar, the Rad Lab put the pedal to the metal and turned out dozens of new innovative gadgets, finally overcoming the wariness of the military. Now working closely together, the two organizations developed tactical devices that became highly successful in the field.

THE RAD LAB EXPERIMENTS WITH NUCLEAR FISSION

Along with microwave radar, the Rad Lab produced other interesting theories, among them was that of nuclear fission. In *Tuxedo Park*, author Jennet Conant wrote:

> *In a sense, the Rad Lab was a catalyst for the burst of creativity and inventive effort that would propel American scientists toward their pioneering achievement in Los Alamos. In the early days of the war, it was* [Alfred] *Loomis, in his role as scientific agitator, who had been the primary force in organizing the country's nuclear physicists to work on radar, at a time when the atom splitters had little to do and fission's useful applications still seemed remote. So by the fall of 1942, when* [Vannevar] *Bush,* [James] *Conant, and General Leslie Groves took steps to form the highly secret atomic bomb development program which was then known as the Manhattan Engineering District (later as the Manhattan Project), along with an urgent effort to develop the component elements in sufficient quantity, they had to look no further than Loomis' Rad Lab for a readily available pool of brilliant minds to draw on.*

1942 BOCA RATON AIR FIELD FLYING OPERATIONS

On February 4, 1942, flying operations moved from Scott Field, Illinois, to Morrison Field in West Palm Beach. First Lieutenant Frank W. Jarek, accompanied by ten pilots, made up what was known as the Flight Section.

Lacking administrative personnel and adequate ground technicians, pilots were assigned additional duties of engineering, supply armament, operations and communications. By necessity, the flying officers worked regularly as "grease monkeys" and "pencil pushers" as well as pilots.

With the completion of runways at the Boca Raton Army Air Field in late 1942, flight operations moved from Morrison Field to Boca Raton. In the *1947 Army Air Forces Training Command Year Book*, Base Commander Colonel Rosenham Beam wrote: "Shortages of equipment and trained personnel plagued the school in its early days. For example until January 1943, only ten pilots were available to fly dilapidated English Hudson patrol bombers on training missions, and often fatigued pilots would compensate for the shortage of crew members and mechanics. The school operated literally 'on a wing and a prayer.'"

CLOSING OUT 1942

In 1942, the soldiers arriving at the Boca Raton Club and the new army post jostled awake the sleepy town of Boca Raton. Yet even with the installation's humble

An English Hudson, the first type of aircraft to arrive at the Boca Raton Army Air Field, crashes on the tarmac. *Photo courtesy of the Boca Raton Historical Society.*

beginnings, residents could see things were about to take a dramatic turn from their laid-back lifestyles.

Planes began to streak across the sky, bringing with them thunderous roars. Hundreds of troops could be seen coming and going from the train and bus depots, and traffic along U.S. Route 1 and Dixie Highway the only north/south routes, increased dramatically. Army uniforms showed up in local watering holes, and civilian companies received large orders for products and services to support operations on the base. Boca Raton had not seen this much action since the land boom of the 1920s.

What Mayor J.C. Mitchell envisioned with his hopeful trip to Washington, D.C., in 1941 was now a bold reality. But this was just the beginning of things to come for the small town of Boca Raton.

1943 TIME CAPSULE

Singer Frank Sinatra debuts on radio's *Your Hit Parade* • Postmaster General Frank C. Walker invents the Postal Zone System • The first withholding of tax from U.S. paychecks occurs • President Franklin D. Roosevelt appoints General Dwight D. Eisenhower as supreme commander of the Allied forces

WORLD NEWS

• Italy's Prime Minister Benito Mussolini and his officials resign; he is imprisoned.
• Soviets make gains in recapturing their territory, ending the siege of Leningrad.
• General Eisenhower and Italian General Marshal Pietro Badoglio sign an armistice.
• Iran declares war on Germany and joins the League of Nations.
• President Roosevelt visits Casablanca, Morocco, and becomes the first U.S. president to visit a foreign country in wartime.
• The Japanese government in Java limits the sale and use of motorcars.

NATIONAL NEWS

• War contractors issue a directive prohibiting racial discrimination.
• The Jefferson Memorial is completed.
• President Roosevelt calls for a $100 billion military budget.
• Canned food and shoes are rationed in the United States.
• The famed botanist, George Washington Carver, passes away.
• The Pentagon, the world's largest office building, is completed at a cost of $64 million.
• The American Broadcasting Company (ABC) is formed.
• The United Negro College Fund is established.

SPORTS

• World Series champion—New York Yankees • U.S. Open golf champion—Was not held in 1943 • Pro football champion—Chicago Bears • Indianapolis 500 winner-Was not held in 1943 • Stanley Cup winner—Detroit Red Wings • NCAA basketball champion—Wyoming • College football champion—Notre Dame • Heisman Trophy winner—Angelo Bertelli from Notre Dame

MUSIC

• "As Time Goes By"—Rudy Vallee
• "Brazil"—Xavier Cugat
• "Dearly Beloved"—Glenn Miller
• "Don't Get Around Much Anymore"—The Ink Spots
• "I've Heard That Song Before"—Harry James
• "In the Mood"—Glenn Miller
• "It Can't Be Wrong"—Dick Haymes
• "Let's Get Lost"—Vaughn Monroe
• "Paper Doll"—Mills Brothers
• "Pistol Packin' Mama"—Al Dexter

MOVIES

• *Casablanca*—Academy Award winner
• *Shadow of a Doubt*
• *For Whom the Bell Tolls*
• *Heaven Can Wait*
• *Phantom of the Opera*
• *Frankenstein Meets the Wolf Man*

COST OF LIVING

New house	$3,600.00
Average income	$2,041.00 per year
New car	$900.00
Average rent	$40.00 per month
Tuition at Harvard University	$ 420.00 per year
Movie ticket	$.35 each
Gasoline	$.15 per gallon
U.S. postage stamp	$.03 each

Food

Granulated sugar	$.75 for 10 pounds
Vitamin D milk	$.62 per gallon
Ground coffee	$.46 per pound
Eggs	$.21 per dozen
Fresh baked bread	$.10 per loaf

1943

CADETS AT THE BOCA RATON CLUB

By the beginning of 1943, construction at the Boca Raton Army Air Field had slowed to a crawl; the only buildings left on the drawing board were officers' quarters and an officers' club. While these buildings were being constructed, training and housing of officers and cadets continued at the Boca Raton Club.

From across America, cadets arrived weekly at the historic resort to attend Officer's Candidate School. So too did instructors, eager to provide newly acquired, top-secret radar instruction to keen young minds.

Porter Richardson was one of an initial thirty instructors fresh out of the Massachusetts Institute of Technology who taught at the Boca Raton Club. Later he was sent to the Aleutian Islands in Alaska where he installed the first airborne radar units. "The only saving grace for our little band of radar persons was that our project was so hush-hush the squadron commanders were told to clothe, feed and house us and let us do as we damned well pleased," Richardson wrote.

One cadet finding his way to the Boca Raton Club in the spring of 1943 was Gerald W. Jones. He spent ten weeks at the facility being hazed as an underclassman and another couple of weeks in the reverse roll.

His first recollection of Boca Raton was arriving at the train depot and asking directions from the stationmaster. Jones, who is white, said the stationmaster "gave me a dirty look because I was at the colored window."

Eventually Jones found his way to the Boca Raton Club a few miles to the east where he bunked in a room with three other cadets. He recalled: "Our days were filled with going to classes and studying for officer training. We spent considerable time learning about close-order drill, and we spent time in the swimming pool learning how to survive if we were on a troop ship and were dumped in the water. On one trip, we

Alvin Thiele trained as a cadet at the Boca Raton Club in 1943. *Photo courtesy of Alvin Thiele.*

went for a swim in the ocean, about an hour's march from our facilities. We spent many hours in classes on the lawn under the palm trees."

Life as an aviation cadet was demanding. Academic classes taught engineering, communications and aerodynamics. Students were confined almost totally to the grounds with scholastic and military training that sometimes lasted between eighteen and twenty hours a day.

Ralph Shulman recalled some of the stringent drills the troops underwent: "I remember standing at attention outside on the grounds in the heat of the summer with gnats covering my eyes. There were so many, I couldn't even see but we weren't allowed to blink."

Seven-year-old Peter Barrett and his best friend, Buddy Lamont, loved to sneak over to the Boca Raton Club and wriggle under the barbed-wire-topped chain-link fence to watch the cadets train. He recalled: "Every Sunday, the cadets held drills on the golf course on the southwest aspect of the club. We often attended as they marched in review and then stood at attention for very long periods of time. The Florida heat and humidity took its toll, and men often fainted. Our 'sport' was to count the number that fainted each week. We were told that if anyone fainted, they were 'washed out' of the air corps, and that if soldiers standing next to them attempted to help, those cadets were also washed out."

"This was a mental, physical and intellectual 'shredding out' for exhaustive assignments to come later when we were commissioned," said Alvin Thiele, a cadet who trained at the club in 1943.

When cadets finished their training at the Boca Raton Club, they were sent to Yale University where they became immersed in a highly academic curriculum. Upon completion of their studies, they received a commission as second lieutenant.

First arrivals at the Boca Raton Club viewed their accommodations as "elegant," as they enjoyed many of the club's amenities, including the swimming pool, golf course and cabanas. Arriving just a few months later, others encountered dramatic changes.

Louis Mikunda arrived at the club in the summer of 1943 to discover all that glittered was not gold. He wrote: "There were over 2000 Air Force Cadets in residence at a time and the conditions were anything but 'elegant.' The rooms had been stripped of carpets and furnishings and contained eight bunks to a room (Green G.I. bunk beds, 4 uppers, 4 lowers). When PT [physical training] ended at 3:30 p.m. and we were to clean up from the obstacle courses and sand football games to dress up for Retreat at 4:00 p.m., there was no water pressure above the second floor—all faucets turned on simultaneously. So we filled the tubs before PT and after PT took turns bathing in the same water. Turns were decided by lottery, and #8 also had to clean the bathroom for possible white-glove inspection during Retreat or dinner. The swimming pool was boarded over. All the grounds and part of the golf course had tents pitched on them where incoming trainees spent their first 2-3 weeks."

John Mangrum also arrived at the club in 1943. With a contingent of other cadets, he was assigned to a room in the back of the hotel. The third-floor former servant's quarters had only one shower. When called for revelry, taking the stairs was mandatory even though an elevator was nearby. And at retreat, they fell in and marched to an area just east of the hotel to the "Colonel Bogey's March" (the theme music from the movie *Bridge Over the River Kwai*).

While most days were filled with repetitive drills and training, Mangrum recalled an especially harrowing event the day commanding officers ordered the men to form combat teams and traverse the canal. Said Mangrum: "They tied a great big rope across the canal, and we were supposed to go across it in full combat gear. Just before I got ready to go, the piling the rope was tied to toppled, and all the men went into the canal. Several of us jumped in to help get them out. Everyone got out, except one black cadet who drowned. No one had bothered to find out that he couldn't swim. They found his body the next day on the beach."

The cadet's workweek required strict adherence to training schedules, but on Saturday nights, the men were free to enjoy whatever entertainment was available. Many took off to Miami Beach or West Palm Beach, while others stayed at the hotel. Those who remained behind enjoyed the sounds of Bach or Beethoven as the Reverend Hugh Savage Clark brought out loud speakers and set up his own concerts. "We sat on the ground on blankets, and he played classical music on the record player," said Mangrum.

In addition to the classical concerts and trips off the field, troops enjoyed patronizing one of Boca Raton's local watering holes, Brown's Restaurant. Said Mangrum: "They had the best cheeseburgers and beer. I remember the jukebox and Betty Davis singing 'They're Either Too Young or Too Old.'"

In the later part of 1943, Hugh Ganser arrived at the Boca Raton Club for an eighteen-week course. With training to include an additional two weeks of instruction on the new radar-countermeasure (jamming) equipment class, new curriculum needed to be developed. Ganser was one of four enlisted men commissioned to develop the curriculum. "The equipment was classified top secret. It was so new, we had to work from prototypes sent down from MIT, and we had no tech-orders to work from," he said. After a few weeks, the course moved into one of the H buildings in the northern section of the base. The structure came complete with a security fence, guardhouse and full-time guard.

During his teaching time at the Boca Raton Club, Ganser recalled two unusual events. The first was an officer who brought bait and line to class to fish from one of the open windows—"with no success," noted Ganser. And the other was his invitation to eat at the officer's mess, an extremely rare summons because enlisted men didn't eat or socialize with officers, on or off duty.

TUSKEGEE AIRMEN TRAIN AT THE BOCA RATON CLUB

Among those who arrived for officer training at the Boca Raton Club in 1943 was an African American from Las Cruces, New Mexico, by the name of James B. Williams. By the time he arrived, Williams had attended segregated grade and high schools and studied pre-med at New Mexico State College. Teaching in Clovis, New Mexico, in 1942, he received word he had been drafted.

Initially assigned to the medical corps at Camp Pickett, Virginia, because of his pre-med background, he was subsequently selected to attend Medical Administrative Officer's Candidate School. But Williams dreamed of becoming a pilot, so he went to the Pentagon and asked to be transferred to the Army Air Corps. Expecting to go to Tuskegee, Alabama, for flight training, instead he was appointed an aviation cadet and sent to Boca Raton. He spent three months in basic training at the facility.

Of his time in Boca Raton, Williams recalled: "Basic training was the same for us as it was for any soldier entering the army. You had to do a lot of drills and underwent hazing by upperclassmen. Food was served army style and managed by self-serve."

One of William's classmates was William (Bill) D. Phears, a farm boy from Ouachita County, Arkansas. He moved to Detroit, Michigan, as an infant, and as a young man, he found work at the Ford Motor Company while attending college at the Detroit Institute of Technology. He enlisted in the Air Corps Reserve in December 1942, and on April 26, 1943 shipped out via train to Boca Raton. In his book, *Ain't, But It Can Be,* Phears chronicles his arrival: "We stepped off the train onto the graveled ditch. I saw

James B. Williams was one of the first black cadets to train at the Boca Raton Club in 1943. He later became one of the Tuskegee Airmen. *Photo courtesy of Creighton University.*

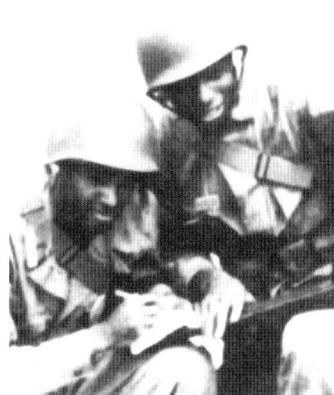

These photos show some of the black cadets who trained at the Boca Raton Club in 1943 with William Phears (top right). They went on to become Tuskegee Airmen, trained fighter pilots for the famous Ninety-ninth Fighter Squadron, slated for combat duty in North Africa. Additional pilots were assigned to the 332nd Fighter Group that flew combat mission from bases in Italy along with the Ninety-ninth Squadron. *Photo courtesy of Jo Addy.*

nothing but trees, and needless to say I had some concern. When the train moved we could see the station, it was about the size of a two-car garage but there were the signs reading 'White' and 'Colored' waiting rooms, with the ticket office in the middle.

"After waiting a short time, a six-by-six Army truck transported us about three miles to a palatial hotel. After disembarking, we were told by an upper class Cadet to, 'turn our backs to the hotel and walk backwards.'"

Once inside Phears was assigned to a large room where he bunked with a dozen other black recruits. Then the upper-class cadets "proceeded to scare the hell out of us, don't do this, don't do that, do this, do that, and what we had to remember," wrote Phears.

While at the club, Phears and the other black cadets enjoyed competitive sports between the squadrons, and he played catcher on his squadron's softball team. After completing basic training, Phears shipped out to Yale University where he received his commission. He was then assigned to Tuskegee Army Air Field.

Of his time at Yale, Phears wrote: "During the two years of the school's [Yale's] existence, 12,830 students entered this non-flying Cadet Advanced Technical Training Program. As a part of the Tuskegee Airmen Experiment, it should be noted that among these 12,830 cadets there were 28 black cadets with never more than three in a class, most of whom went through the Boca Raton basic training."

Phears noted the following cadets who were stationed with him in Boca Raton and went on to become Tuskegee Airmen: Charles O. Southern (retired colonel), Matthew Merriweather, Alfred Scott, William A. Kelley (retired lieutenant colonel), James D. Solomon (MD, PhD), Frederick H. Williams, Robert Randall, Raymond K. Dewberry (retired lieutenant colonel), Arthur Saunders, Alfronzo Dowel and Luna Mishoe (PhD and college president).

PILOTS ASSIGNED TO THE BOCA RATON ARMY AIR FIELD

The Boca Raton Club wasn't the only U.S. facility commissioned to train cadets; they trained virtually all over the country. Once commissioned, many of them reported to the Boca Raton Army Air Field for further training and flight time.

Nelson C. Godett, a flight officer and bombardier, transferred to the base in 1943. He recalled: "I was enrolled in the Air Observer School, learning how to bomb from B-24s with radar attached to the Norden bombsight. We attended school two days per week and flew practice bombing missions three days per week, sometimes vice versa. Flying missions took us to Jacksonville, Tampa, Fort Myers and back to the Boca Raton Army Air Field. Other flights were to Bimini (an island in the Bahamas). All these flights were for practicing bombing by radar."

(The Norden bombsight became one of the most important military secrets during wartime. The device included an automatic pilot and used a mechanical analog computer to determine the exact moment a bomb should be dropped to accurately hit its target. At the time, it was thought to be the key to daylight strategic bombing.)

Captain Harry Fromme, squadron commander (on left, first one standing), poses with the initial pilots assigned to the Boca Raton Army Air Field. In the background is a B-24. It is believed the two white lines on the nose were made by the photo lab, marking out the number of the aircraft for security reasons. *Photo courtesy of Lt. Col. Manuel Chavez USAF (Ret.).*

Pilots flew coastal patrol, scanning for U-boats. On one such run, Peter Wickert—serving as crew chief, a position that supervised all operations onboard the aircraft and all ground mechanical repairs—was aboard a B-26 that had to be ditched off the Florida coast. "We ran out of fuel. It was quite a thrill jumping out of one of those planes as the hole you had to jump through was only about three feet square. The crew spent three hours in the water before being rescued," Wickert said.

While gasoline was rationed in the civilian sector, on occasion the Boca Raton Army Air Field had more than it needed. "There was so much gasoline," Wickert recalled, "that it was dumped on the ground at the end of the flight line and burned."

Manny Chavez arrived in Boca Raton in early 1943. After living at the Boca Raton Club for two months, he moved to barracks next to the base hospital where he had his own small room. Half of the building was used for housing, and the other half was converted into a day room, consisting of a small bar with a part-time GI bartender, a pool table, a nickelodeon, a dartboard and six square tables with chairs.

Lieutenant Manuel "Manny" Chavez (front row, left) gives instructions to his flight crew. *Photo courtesy of Lt. Col. Manuel Chavez USAF (Ret.).*

Chavez was among twenty-five young pilots fresh out of flight school who were assigned to training missions in a variety of aircraft. They would fly parallel to the coast in order to practice locking on to buoys, small boats and ships in the Gulf Stream.

At that time, the Boca Raton Army Air Field had few aircraft, but as planes began to roll off assembly lines, the airfield received surplus aircraft from other training fields. Then, pilots were in short supply. Said Chavez, "I was sent to Love Field in Dallas, Texas, to ferry a B-34 back to Boca with two good sergeants but no co-pilot."

When Chavez transferred to the Boca Raton Army Air Field, Colonel N.L. Cote was base commander. According to Chavez, Cote was touchy about his unusual name, Narcissus Lilli Cote and threatened troops with death if his complete name appeared on any order or document. One day the daily order came out marked "By order of Cpl. Cote"; corporal is a low enlisted man's rank. Whether it was a typo made easily by hitting the "p" instead of the "o" on the typewriter or a deliberate act, Chavez recalled, "Talk about a furious recall of all copies. The typist was immediately transferred to a combat zone."

Colonel Narcissus Lilli Cote (center front) escorts Major General Walter R. Weaver and other top brass on an inspection tour of the base. This rare photo of a radar training building, complete with guard and guard shack and surrounded by a barbed-wire fence, is considered one of the most significant photos of the former base. Through the T-number on the guardhouse, historians were able to establish this building as Mizner Oaks Apartments, one of the few remaining off-campus structures. As this book went to print, this building was in the process of being sold to make room for a condominium complex. *Photo courtesy of the Boca Raton Historical Society.*

Hugh Ganser served on the base during Cote's command and also recalled the infamous base commander:

> *The new commanding officer of the base was a full colonel with an apparent mental problem. He changed all the house rules of the base to make life as miserable as possible. First he stopped the camp transportation buses at the base gate instead of allowing them to drive the half-mile into Boca. Then he stopped the buses at a turnaround a half-mile inside the gate. The next order required that all personnel going to chapel or Mass on Sunday must march there in formation. Our gutsy chaplain, Capt.*

Murphy, took the train to Washington to register a complaint. He returned a major, and the company commander got his overseas orders within a week.

HUSH, HUSH

The Boca Raton Army Air Field served as the Army Air Corps' only airborne radar-training facility during wartime. As such, the facility and training done there were considered top secret. According to Wesley Frank Craven and James Lea Cate in their 1949 publication, *The Army Air Forces in World War II*, vol. 6, "Men and Planes,"

> *Location of targets and bombing by electronic devices was one of the most significant developments in air war, BTO* [bombing through overcast] *training was initiated in October 1943 at Boca Raton Army Air Field, Florida and within a year it became the dominant course of the radar program.*

Before arriving at Boca, troops who were selected for the top-secret radar school underwent in-depth testing and training at installations in the Midwest. Bill Robison, a radar trainee, explained what kind of training the men had:

> *The significance of this training mission* [at the Boca Raton Army Air Field, or BRAAF] *and the excellence with which it was planned and executed should be understood and appreciated. The great majority of these mostly young trainees had no previous technical training in electronics. Some had been called to active duty before they completed their high school education. Virtually no one at the time had television. Some did not have radio, and some came from socioeconomic circumstances wherein they did not have a telephone while growing up. Thus, the majority did not bring life experiences that equipped them with some intuitive understanding of the technology which they were about to be trained.*
>
> *It is important to note that the planners/educators were not content to teach in a rote-like manner the sequence of "cook book" procedures that would enable the radar mechanic to somehow fumble his way through the maintenance of the equipment. Thus, before attending radar school at BRAAF, the radar trainees were required to satisfactorily complete a sixteen-week radio mechanics course at Truax Army Air Field at Madison, Wisconsin. Thereafter, they were required to successfully complete a pre-radar electronics course at Chanute Field at Rontoul, Illinois This was a college level electronics course wherein the trainees learned the theory and functions of the circuits that were utilized in the radar used for training at BRAAF. There was rigorous testing along the way, and those that failed to maintain satisfactory grades were "washed out" and sent to gunnery school.*

United States Army

Army Air Forces Technical School

Be it known that

Pvt Harry Magafan   33193728

has satisfactorily completed the course for

Radio Operators

as prescribed by the Army Air Forces Technical Training Command
and given at Madison Field, Wis.
In testimony whereof and by virtue of vested authority
I do confer upon him this

=====**Diploma**=====

Given on this   19th   day of   December   in the
year of our Lord one thousand nine hundred and forty-two

Colonel, AC.
Commanding

After finishing radio operator's school at Army Air Forces Technical Training Center at Madison Field, Wisconsin, Harry Magafan traveled south to attend radar school at the Boca Raton Army Air Field. *Photo courtesy of Harry Magafan.*

Harry Magafan arrived at the Boca Raton Army Air Field in 1943. He poses in his flight gear. *Photo courtesy of Harry Magafan.*

New lines of offensive and defensive electronics rolled off benches at the Rad Lab and eventually found their way to the Boca Raton Army Air Field where they became the foundation of the highly secretive radar-technical-training program. Radar training was held in H-shaped concrete buildings scattered throughout the base. High fencing topped with barbed wire surrounded the structures, and each had a guard shack positioned at the entrance. Classrooms occupied the vertical legs, and the horizontal bar contained offices, electronics laboratories and latrines.

Instruction lasted as few as five weeks and as many as thirteen months, depending on the equipment and training requirements. Instructors came from Massachusetts Institute of Technology, Sioux Falls Technical Radio School and other training facilities to teach classes six days a week in three 6- or 8-hour shifts each day. Divided into two sections, classes allowed for 1 hour of classroom work followed by 1 hour of laboratory work. During the laboratory session, the men put theories from the classroom into practice. The men took 10-minute breaks every hour and marched to and from the mess hall.

Between eight and ten men attended each class, and several classes went on in a building simultaneously. Everything at the school had to be learned in the classroom by watching the instructor who used a blackboard and models. To be caught with a piece of paper or a pencil—or trying to write anything down—was a court-martial offense.

Bob Davey, who attended radar school, explained the guarded nature of the instruction: "Everything was secretive. We were warned that discussion of even the word 'radar' off the base was an automatic court-martial. When we came into the building for class a few minutes before noon, let's say, the other class was leaving out the back door while we were allowed in the front door. When your name was called, you answered with the last four digits of your serial number to be checked off and then you were checked off at the end of the six hours going out the back door the same way."

During the summer months, the physical conditions under which trainees attended school were difficult. Bill Robison recalled: "There was no air conditioning. To make matters worse, the trainees were required to wear full fatigue dress (no shorts), including laced up leggings and G.I. boots. Needless to say, during the summer months, the trainees were dripping wet throughout the training period."

Ed Malavarca, who was assigned to Ground Control Approach, recalled that the courses were "pretty rough" and that the army "tried to squeeze everything into a short period of time."

Trained as an airborne radio operator, Gerald Weiss came to the Boca Raton Army Air Field from Scott Field, Illinois. His course was five weeks long and included flights on B-17s on antisubmarine patrol. He recalled: "Since I arrived in July at the height of the summer, it was extremely hot and humid, and we had to wear our flight gloves when we got into the aircraft to avoid burning our hands on the controls."

Another way to avoid getting burned was to make a passing grade. Philip Kantor remarked on the pressure put on the troops: "I attended school there studying Ground

Control Approach (GCA), which was the fairly new equivalent of a mobile unit near a landing strip bringing planes in safely for landings. We were clearly told that our being there was contingent on our maintaining a minimal passing average, which I believe was 3.0. Failure to maintain this minimal average was an immediate dismissal from the program and transfer to the infantry."

Some troops unfortunate enough to be assigned a 6:00 a.m. to noon class found it difficult to stay awake. Ed Wallish was one of the men attending class at this hour. He recalled: "The first minutes in class were highlighted by a constant struggle to stay awake. Agitated instructors would yell, throw erasers and prod sleeping students awake. Sometimes if you felt you were dozing off, it helped to stand up in class. Between classes, a trip to the latrine for a face-full douse of tepid water helped. The best solution came after the first four weeks, when we moved to a new building and began the afternoon shift from noon to 6:00 p.m."

For the next twelve weeks, Wallish studied basic electricity, resistors, capacitors, inductances, schematics, power supplies, electronic tubes, electronic tube circuits, radar timing circuits, magnetron oscillators, modulators and antennas.

The radar-countermeasure school was one of a number of advanced schools that trainees attended. The purpose of radar countermeasure was to obtain intelligence about enemy radio and radar transmissions in order to jam and confuse enemy communications and radars. The equipment studied was:

> • *AN/APR-4: A radar receiver for searching frequencies from 60 MHz to 6,000 MHz. The receiver featured changeable tuning units to change frequency bands. The tuning units used motorized scanning to sweep the frequencies.*
> • *AN/APR-5: A radar receiver to search frequencies in the ten-centimeter band. The receiver used a tuned cavity to obtain the proper frequency.*
> • *AN/APT-5: A barrage jammer that transmitted noise signals to jam enemy radar. Electrons colliding with gas molecules in a gas-filled tube generated the noise.*
> • *AN/APA-11: A signal analyzer used in conjunction with an AN/APR-4 to obtain signal intelligence about radar transmissions.*
> • *AN/APA-17: A signal analyzer used to find the locations of enemy radars. The transmissions were received on an AN/APR-4 by way of a rotating receiving antenna and displayed on a circular scanned cathode ray tube (CRT). Locations were mapped by vectoring for the greatest signal strength on the CRT display.*
> • *AN/APQ-10: A spot jammer for the ten-centimeter band. It featured an AOT-5 variant coupled to a panoramic adapter. Coordination with the jamming transmitter was made via a system of color-coded pluggable RF [radio frequency] transmitter units with tuneable magnetron tubes. The tube filaments for the pluggable units were supplied with power from a special powered unit rack so that time was not lost by waiting for the tube filaments to heat up.*

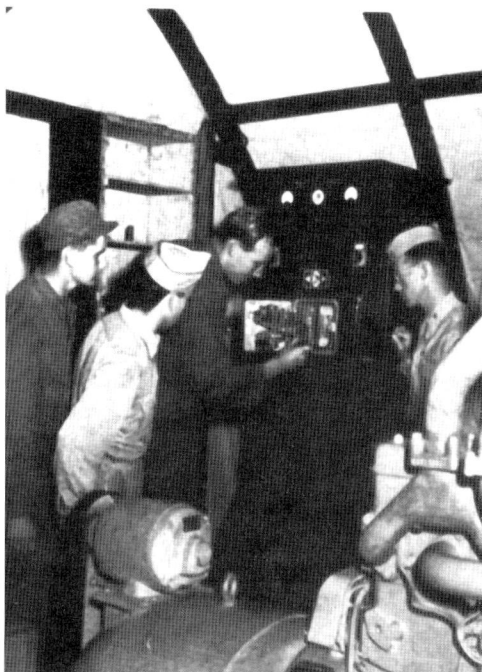

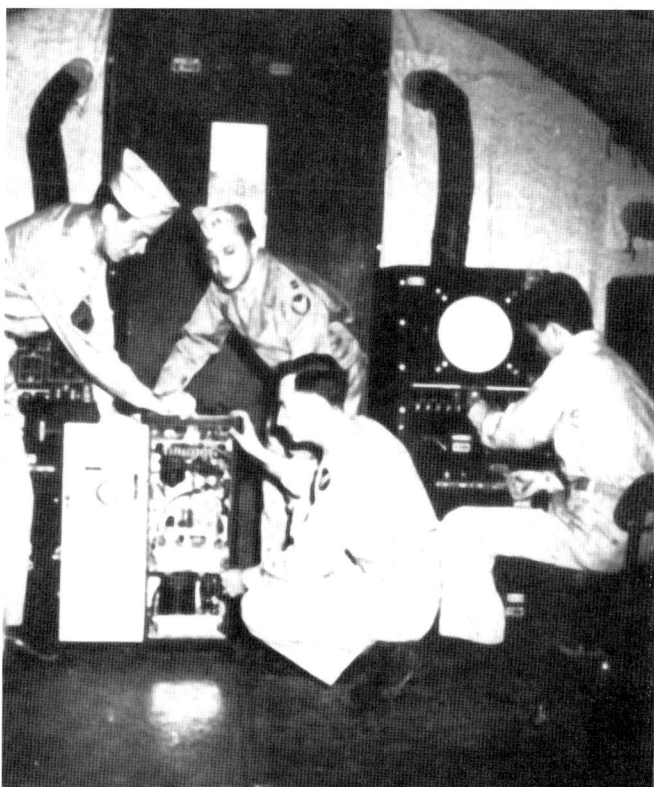

These photos depict some of the radar training undergone by troops attending the Army Air Corps Technical School of Radar at the Boca Raton Army Air Field. *Photo courtesy of the Boca Raton Historical Society.*

Philip Kantor studied Ground Control Approach, which was used to assist in the safe landing of aircraft. *Photo courtesy of Philip Kantor.*

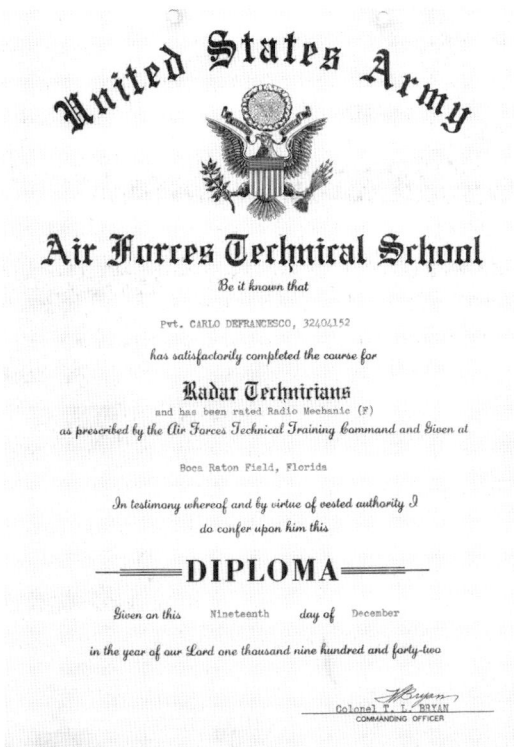

Private Carlo DeFrancesco finished radar-technician classes and was awarded his diploma. It is signed by Commanding Officer Colonel Thomas Bryan. *Photo courtesy of the Boca Raton Historical Society.*

Once troops finished classroom training, they transferred to flying classrooms where they put their ground training into practice. B-17 bombers, fitted as airborne-radar classrooms, were flown over the Atlantic from Boca Raton field. "We practiced dropping sand filled bombs on a rock called Little Isaac that stuck up out of the ocean," said Richard Gerber.

Paul Metro remembered his flight aboard a B-34 Lockheed Ventura, a two-engine bomber: "My first airplane ride ever was a radar-training mission. This was an experience because the radar didn't work, and it turned into a roller-coaster ride over the Florida Keys. I got kind of queasy in the stomach, but didn't get airsick. The pilot was noted for giving trainees that kind of 'fun.'"

Many of the men trained as mechanics and became part of the ground crew. Their job was to install and repair radar and communications equipment in the airplanes. Joe Ornstein, who served on the base in the summer of 1943, spent his practical training with P-47 Thunderbolts. He changed the kilocycles on pilot communication devices so the Germans couldn't pick up their signals, and he installed dynamite under the radio. "If the plane crashed, the wires would burn and the Germans could not duplicate the radio," Ornstein said.

GROUND CONTROL APPROACH (GCA)

One of the most important radar devices developed and tested at the Boca Raton Army Air Field was Ground Control Approach (GCA). This apparatus was designed to track planes and assist them in landing. Manny Chavez was called upon to assist in the testing of this new device. He later became the first pilot to experience its true capabilities.

One morning, when I was not scheduled to fly, a civilian showed up at Operations to request that an aircraft assist him in an experiment with a new radar system. He was an electronic scientist working for the government, whom we later referred to as the guy with the long hair and thick glasses. He was British. I was assigned the mission in the B-18 to make landing approaches. He explained that I should fly about 3,000 feet, five to ten miles from the landing field and contact him by radio on a specific frequency. He said that he would then give me compass headings to fly, align me with the landing runway, and tell me when to start my descent for a landing.

His equipment was parked in a two-wheel trailer with a tarpaulin cover in the center of the landing field just off the east/west runway. At 500 feet above the runway I was to abort the landing and make another run. This was the beginning of GCA, Ground Control Approach. I continued flying for this scientist for several days and other pilots also were assigned to do practice approaches and the results were impressive. It was a major contribution to the war effort, specifically for the successful air war in Europe.

In those early days of flying, navigational aids were primitive compared to today. We only had radio direction finders (RDF) for landings and approaches. These radio compasses were our principal cross-country navigational aids. Flashing lights and Morse Code signals identified airfields.

One afternoon I was returning from a radar training mission over the Bahamas. It was in February, the dry season in Florida, and there were numerous grass fires in the Everglades. The smoke along the coast was so thick that it was almost zero visibility. The Boca Raton Field RDF (radio direction finder) was not functioning, so I couldn't get a direct bearing to return to the base. I was in the middle of the Bermuda Triangle (also known as Devil's Triangle), and wanted to get home. Remembering my friend with the long hair and thick glasses, I called him on his frequency, praying that he would be at his set under the canopy. His answer came back loud and clear, "Roger Dodger son, I have a good fix on you. You are 19 miles southeast of the field, take a heading of 287 degrees and maintain 2,500 feet." He expertly took me west of the field, then lined me up with the 09 (east) runway, and guided my descent to 200 feet over the end of the runway, right on the money. It was his and my first radar approach under adverse weather conditions, with God as my co-pilot.

The development and testing of the GCA was a significant accomplishment for the obscure Boca Raton Army Air Field, and it would later play an important role in the cold war.

From June 24, 1948, to September 30, 1949, GCA played a vital role in the coordination of planes into and out of Berlin during the Berlin Airlift. At this time, the Soviet Union set up a blockade around the city and refused to allow ground transportation to deliver supplies.

The United States and Great Britain flew C54s and C47s, respectively, into Berlin to deliver lifesaving food, fuel and supplies. At peak strength, 255 C54s landed in Berlin in what became known as the Easter Parade in which 1,398 sorties (one landing in Berlin every minute) brought in 12,940 short tons of supplies. During the airlift, the United States, Britain and France flew 278,228 sorties into Berlin, airlifted 2,326,406 short tons of food, coal and supplies and transported ten thousand people—all with the assistance of GCA.

While GCA was used extensively throughout the military from the 1940s through the 1960s, Instrument Landing Systems gradually replaced it.

CHEMICAL WARFARE AND FIREARMS TRAINING

In addition to radar instruction, troops received other types of training, including chemical-warfare and firearms training. In July 1943, Lieutenant George P. Dawson became the post chemical warfare officer. His mission was to train all personnel to protect themselves in the event of a chemical attack and to educate troops in the types of chemicals used in chemical warfare. To that end, troops were subjected to regularly scheduled chemical assaults for which they donned gas masks. On one such occasion,

Troops undergo chemical-warfare training. *Photo courtesy of the Boca Raton Historical Society.*

Other wartime training included regular trips to the firing range. Skeet shooting was initiated as part of pilots' regular training to help them prepare to lead their targets in air combat. *Photo courtesy of the Boca Raton Historical Society.*

Members of the Women's Army Auxiliary Corps (also known as WAACs) stand at attention as they undergo inspection by the base commander. *Photo courtesy of the Boca Raton Historical Society.*

Paul Metro noted that he "couldn't get his mask on quick enough and coughed for hours afterwards."

WOMEN ARRIVE ON BASE

Not all 100,000 troops who passed through the Boca Raton Army Air Field between 1942 and 1947 were men. In fact on May 20, 1943, the field got its first contingent of WAACs (Women's Army Auxiliary Corps): 125 enlisted members and 2 officers. Second Officer Vera M. Corlett was appointed the first commanding officer of the newly designated 750th WAAC Post Headquarters Company AAF (Army Air Force).

Boca Raton Radar School describes what it was like to be a WAAC at Boca Raton:

> *The newly arrived WAAC's were forced to endure an extremely rugged existence. At their former training stations their hopes had been augmented to breathless heights by tales of bands, parades, and good cheer with which other posts had received their first WAAC's. All the tact and kindly but forceful discipline of an exceptional officer was required to smooth the ruffled feelings of the first contingents to arrive at Boca Raton Field. No bands played, no transportation met*

them at the station, and no savory odors greeted them from the mess hall; in fact, there was no segregated WAAC mess at all, and very little besides bare walls, a roof, and a floor served as a barracks. A three-week restriction to this ultra-rustic area proved to be the final straw.

In response, the first few contingents, under the supervision of Staff Leader Caroline Rude, cleaned the barracks and day room, moved in furniture and policed the area. They painted the walls, upholstered the furniture and put up curtains. As a result, the women spent their free time in much more pleasant surroundings.

Volunteer details acted as firemen, and two weeks later, when the area mess hall opened, these subsistence details served on kitchen patrol and as cooks. During this period, work had to be performed in men's overalls, as the company supply had not received WAAC fatigues.

After the women proved capable and competent to perform duties that the men considered "beyond their capabilities," more requests for enlisted women poured into headquarters. The change in attitude was reflected in an article in the *Transmitter*: "Many of the soldiers who scoffed at the idea of 'women soldiers' last May and laughingly remarked, 'What the hell, Women can't take it in the Army,' have cheerfully eaten their words—without pepper and salt."

The main objective of the WAACs was to participate in the operations of the base. In addition to working eight hours, the women also took water-safety and chemical-warfare courses. Supervised recreation was injected into their military training and was held in Gym 4 one night a week. It consisted primarily of volleyball and basketball games, and proved to be a morale booster, especially for the younger women.

WAACs served on the post in the following capacities: payroll, finance, general, statistical, post office, special service, PX, orderly room, supply and switchboard clerks; cryptographers; photographers; teletype and radio operators; hospital ward workers; medical lab technicians; cooks; librarians; messengers; plane dispatchers; motor drivers; mechanical and repair workers; and dental assistants.

U.S. Army Nurse Corps

Making their way south to the Boca Raton Army Air Field from the University of Tennessee in January 1943 were three graduating nurses: Bernice Butler, Gladys Gillis and Ann Hooten. Just commissioned as second lieutenants, they arrived at a remote depot west of the base. Gillis recalled: "We arrived at night with no one to meet us. An officer who happened to be riding with us phoned for transportation, and a vehicle was sent to pick us up and take us to the base."

They were each assigned a cot in the nurse's quarters; accommodations that had neither air conditioning nor heat. Orange crates served as bedside tables. Winter uniforms and shoes were issued at the hospital; however, after several weeks on base,

Gladys Gillis and Bernice Butler pose in their nurses uniforms just after graduation at the University of Tennessee. *Photo courtesy of Bernice Butler-Chavez.*

Nurses stationed at the Boca Raton Army Air Field in early 1943 were (from left to right) Joy Barager, Anna Belle Longwater, Edna Schneider, Ruth Reid, Dorothy H. Lucky, Louise Findley (chief nurse), Ann Maguire, Mary O'Conner, Rita McLarnan, Mervin, McIntyre, Ruth Deese, Gladys Gillis, Ann Hooten and Bernice Butler. *Photo courtesy of Bernice Butler-Chavez.*

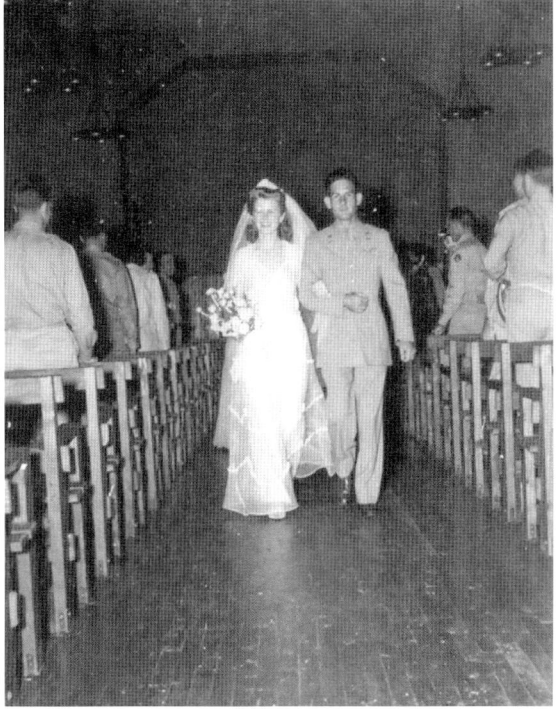

Manny and Bernice Chavez stroll down the isle of a base chapel after nuptials. *Photo courtesy of Lt. Col Manuel Chavez USAF (Ret.) and Bernice Chavez.*

Ann Maguire served as a nurse in the three-hundred-bed base hospital. *Photo courtesy of the Boca Raton Historical Society.*

Butler, Gillis and Hooten traveled by bus to Saks Fifth Avenue on Miami Beach to purchase, at their expense, summer uniforms.

Without any military training, arriving nurses were taught quickly how to march and salute. "We also had calisthenics sessions to keep us in shape, as if walking the hospital corridors twelve hours a day wasn't enough," Butler said.

The hospital had been in operation for only a few months when the three arrived. They found it understaffed by doctors and nurses. Butler recalled: "We were required to work twelve-hour days, six days a week. There were only a few doctors, but they were excellent and had a good relationship with the patients."

The hospital had sulfa drugs but not penicillin or antibiotics. While penicillin was discovered in 1928, it had not been refined, and it wasn't until 1943 that the required clinical trials were performed, showing penicillin to be the most effective antibacterial agent to date. Penicillin production quickly escalated, but initial quantities were shipped overseas to treat wounded Allied soldiers. To stave off infection in patients on the base, doctors prescribed sulfa tablets and sprinkled sulfa powder on wounds and incisions.

Positive-pressure breathing machines, to assist patients with deep breathing, were also nonexistent "We used paper sacks held firmly over the patient's nose and mouth to build up CO_2 which stimulated deep breathing. This was to prevent pneumonia in surgical patients," Gillis recalled.

When the nurses first arrived, they ate in their own mess hall that, according to Butler, had an "excellent cook." But shortly thereafter, they were required to eat at the hospital mess hall where the quality of food was "very different."

As officers, the nurses enjoyed the officer's and cabana clubs at the Boca Raton Club from 1942 to 1944, and later the officer's club on the base. Evening gowns were allowed when attending formal affairs, otherwise military or white nurse's uniforms, with hat, were worn.

Aside from the officer's club, other forms of recreation included riding olive-drab bicycles provided for the nurses. Butler recalled: "One late afternoon, Ann Hooten and I were riding bicycles on the narrow perimeter road that led to flight operations. A car stopped and it was Captain Sharago, the bandleader, who had recently been a patient in the hospital. He introduced me to his passenger, a young pilot, Manuel Chavez, from New Mexico. Lieutenant Chavez invited us and other nurses to join them at the small officer's club in their nearby quarters. We became good friends, often dating together with nurses Hooten and Gillis and their boyfriends."

Romance was alive and well at the Boca Raton Army Air Field, and those dating days gave way to marriages for all three nurses in 1943. Lieutenant Ann Hooten married Lieutenant O.J. Weber in May; Lieutenant Gladys Gills married Lieutenant Clinton Hanson in June; and Lieutenant Bernice Butler married Lieutenant Manuel Chavez in July.

Ann Maguire was one of six nurses initially staffing the three-hundred-bed base hospital. In a 1983 interview for the *Fort Lauderdale News/Sun-Sentinel*, she stated:

Dubbed the "Flying Angel," flight nurse Edith Mize-Jackson flew more than fifty missions near the front lines. *Photo courtesy of the Boca Raton Historical Society.*

"Of course there was gonorrhea and syphilis but injuries were mostly the result of motorcycle accidents. And there were rattlesnakes all over the place."

THE "FLYING ANGEL"

Some of the nurses performed duties as difficult and dangerous as those fighting on the front lines. One such officer was twenty-two-year-old Lieutenant Edith Mize, who stated: "I joined the air force because I wanted to be a flight nurse. When they said, 'You're going to Florida,' I said, 'Great, I've never seen it.'"

While stationed at Boca Raton for seven months during 1942–1943, Mize met fiancé Fred Jackson. After Mize transferred to Kentucky for flight-nurse training, the couple married and a few days later the new bride was transferred overseas. Flown behind enemy lines, her job was to care for the wounded as they were airlifted to hospitals in England and the United States.

Dubbed the "Flying Angel," Mize was assigned to the 816th Medical Evacuation Squadron of the Ninth Air Force. She flew fifty missions near the front line, witnessing such events as Normandy, the Battle of the Bulge, and the biggest event of them all, V-E Day, which fell on her birthday. She also helped evacuate survivors from the death camp at Buchenwald.

In the *Florida Nursing News* dated July 8, 1985, Mize recalled: "I flew alone. There weren't any doctors to help me. I had to evacuate the soldiers who were severely wounded by myself. I was in charge. I made about four trips from Scotland to New York with badly wounded soldiers. I had some pretty close calls. I had to radio ahead sometimes when I didn't think someone was going to make it, but no one ever died in my care, thank God."

On one such mission, Mize witnessed two planes of soldiers collide as they returned to base from a bombing mission. "I didn't have time to cry," she stated.

For her efforts, Mize was awarded the Air Medal, the American Theater Service Medal and the European African Eastern Service Medal.

BOCA RATON EXPERIENCES A GOLD RUSH

Ignited by the arrival of thousands of troops, the economic boom Boca Raton Mayor J.C. Mitchell envisioned turned out to be something for which the small, close-knit community wasn't quite prepared.

In her article "World War II in Boca Raton: the Home Front" in *The Spanish River Papers*, Drollene Brown wrote: "Max Hutkin likened it to the 'gold rush of the Klondike.' Hungry Army personnel soon discovered the Hutkins' store. Max and his wife, Nettie, got up each morning at 3 a.m. to make thousands of sandwiches, which would be sold each day by 2:00 [p.m.], for 15 cents each."

Other locals found their riches in housing, a precious commodity during the war years. While troops inhabited the field's barracks or the Boca Raton Club, wives and dependents found housing wherever they could. Many times that was as far away as Boynton Beach, fifteen miles to the north, or Fort Lauderdale, eighteen miles to the south.

The town's Old Floresta section—one of the neighborhoods designed by Addison Mizner in the 1920s—was pressed into service to house officers. Beach rentals that typically housed seasonal visitors now accommodated military dependents year-round.

Two Federal Public Housing Authority projects were built to house military families. Palmetto Park, a housing project for whites, was a series of attached units located across from the base on the south side of West Palmetto Park Road between the 200 and 300 blocks. Its official name was the Palmetto Park War Housing Project, but local residents referred to it as "The Project." It was later renovated and became known as the Garden Apartments. These buildings were subsequently razed to make way for more contemporary structures.

The other housing authority project was known as Dixie Plaza. Located across the canal from Squadron F, it was a series of small individual homes used to house black families. It is known today as Dixie Manor and sits on the northeast corner of Glades Road and Dixie Highway.

Even with these two government projects, lack of housing became so dire that local residents took in boarders. One such resident was Lillian Race Williams who lived,

WHERE TO LIVE?

LIVE ON THE BEACH AND
ENJOY YOUR STAY WHILE AT
UNITED STATES ARMY TECHNICAL AIR SCHOOL

ROOMS, LODGING AT LOW COST

AVAILABLE AT FOLLOWING PLACES.
ALL MEMBERS OF OCEAN BEACH ASSOCIATION

MAR LAGO ....Mr. and Mrrs. L. A. Robbins
PINES BY THE SEA ....Henry Metcalf
OCEAN SHORE COTTAGES................Mrs. Oliver A. Brown
BRINY BREEZES PARK ....Paul L. Miller
PELICAN BEACH APTS. ....E. E. McIntosh, Mgr.
MRS. H. A. HUNT
DEG HOUSE AND COTTAGES ....Dorothea E. Galvin
OCEAN VIEW APTS. ....George H. Riley
MARINE-LAND TERRACE HOTEL..............J. B. Reid, Mgr.
CORAL COVE APTS. ....E. P. Scherer
SANDOWAY EAST HOTEL
OCEAN TERRACE APTS. ....Allabough & Son
HALIDAY HOUSE ....Capt. Haliday
SEACLIFF COTTAGES ....Mrs. Wm. C. Burton
BLUE WATERS ....Mr. and Mrs. Rudolph Hertwig
BEACH HOUSE INN ....Gladys Dixon
BOCA RATON VILLAS ....Mr. Barrett
T. J. GRAHAM
OCEAN VUE ....Charles T. Rich
HOOBLER'S APTS. ....W. M. Dow, Mgr.
SUNNYSIDE ....Andrews Sisters
MRS. COLSON APTS.
FONTAINE APTS. ....Dr. and Mrs. Fontaine
DUNRAVEN COURT ....J. A. Rainford
OCEANIC ....Marie McCollom
PALM HILL ....Mr. and Mrs. Herbert L. Zipf
EL FARO APTS. ....Mr. and Mrs. C. A. Bock
LIGHTHOUSE COVE ....Mrs. Juanita Gentry Clay
SILVER THATCH INN ....Mrs. M. Rambaut-Hill
KESTER'S COTTAGES ....W. E. Kester
VISTA DEL MAR ....Mr. and Mrs. A. W. Schmidt
PAMENTIER'S BUNGALOWS......Mr. and Mrs. C. Pamentier
HOBART'S APTS. ....Mr. and Mrs.Hobart
CARDINAL COTTAGES ....Mr. and Mrs. Herman Berke
MARINE VILLAS ....Mr. and Mrs. S. E. Northway

SEE LOCATIONS ON MAP

LOOK FOR THIS EMBLEM

OCEAN BEACH ASSOCIATION

AND GET A SQUARE DEAL

OCEAN BEACH ASSOCIATION

BOCA RATON, FLORIDA

"Organized for Patriotic Service"

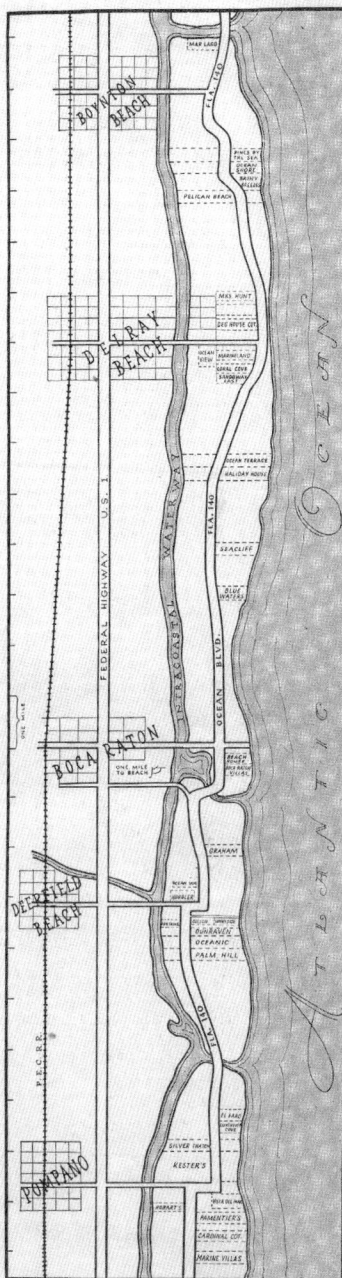

FILE: ...... Boca Raton

A lodging poster directs soldiers to beach accommodations for their families from Boynton Beach to Pompano Beach. *Photo courtesy of the Boca Raton Historical Society.*

with her widowed mother, at 301 SW First Avenue in a small two-bedroom house that her parents purchased in 1917. According to Ella Elizabeth Holst, author of "The Life of a Boca Raton Woman" in the 1986 edition of the *Spanish River Papers*, to make ends meet after the death of her father, Lillian took in boarders, many who were caddies at the Boca Raton Club. But when the season was over, so was her income.

Her economic break came when soldiers poured into the area seeking off-base housing for their wives and families. She soon had more business than she could handle. To make extra rooms, Lillian put up cardboard partitions throughout her home—although she kept a private room for her ill mother until her death.

At one point, she even granted permission for Louis M. Schultze, to build an eight-foot-by-twelve-foot cabin on her property that he would occupy for $5 per month. But even this solution didn't solve the overcrowding. Lillian wrote: "I rent rooms: Also two parts of the garage where there are two mexican [*sic*] families and each have a young baby. Also I have a small cottage with at family in it who rent by the week, since December. The Mexicans have been here since October. There are 25 in all who stay here, 17 in the house. I sleep in a corner of the piazza, and another couple in the opposite side."

Pat Eddinger Jakubek (right) plays with a friend in front of their Palmetto Park unit. *Photo courtesy of Pat Eddinger Jakubek.*

Lillian was chief cook and bottle washer for her boarders. The washing was done in the back yard in a tub with washboards, and cooking was done on a kerosene stove. Her other chores included general housecleaning, yard work and home maintenance. While the work was hard, it did allow her to meet her financial obligations.

She was an outspoken woman who kept meticulous records of her boarders, recording the names and hometown address of each along with their length of stay and amount of payment. She would also write personal comments next to each roomer's name, such as "good riddance," "good boy, but drank," or "kid with big mouth and head."

In 1976, ill and nearly destitute, Lillian sold her land and moved to a nursing home. She donated her beloved home, Singing Pines, to the Boca Raton Historical Society. So named by Lillian because the tall pines surrounding the home caught the breeze and played a melody, Singing Pines now sits on Second Avenue and is the Boca Raton Children's Museum. It is one of the few original Boca Raton homes still standing.

Other Housing

Waino Ray, of Gorham, Maine, arrived at the Boca Raton Club in 1942 where he attended radar school. In 1943, he became a radar instructor at the Boca Raton Army Air Field and brought his wife, Doris, to join him. She wrote:

We lived in "Poinciana" (The "Indiana House") from July 1943-Feb. 1946. We spent Thanksgiving in the house in 1942 with Pvt. Harry and Marjorie Brown and then took over the house when I had finished my teaching year in Conn. And Harry went on to Officer's Candidate School. Since there were two bedrooms in the house, we had couples living with us, also my parents for vacations.

Clara Riggs who owned "Poinciana" and "Crow's Nest" (The "Texas House") was a remarkable woman, an accomplished swimmer, pianist and organist. She had 51 varieties of tropical plants, including all kinds of fruit trees, on the two properties. She knew all about them and tested all the fruits, including the castor bean which made her violently ill.

Crow's Nest was named for Clara's pet crow "Blackie" who lived in a large outdoor cage between the two houses. Blackie could talk and when he called, "Hello, Hello," it startled the soldiers walking to the beach. He was an embarrassment to me if I happened to be in the yard at the same time!

Our daughter, Marcella Ann, called Sally, was born October 10, 1945 at the BRAAF Hospital, a wooden structure resembling the barracks, but with a staff of fine Army doctors and nurses. Families of the soldiers were given excellent care.

The war years in Boca were difficult ones because of the sad and tragic world conditions, but we learned to live one day at a time and to be grateful for the good things, also. We made lasting friendships, collected shells and enjoyed the beach, within walking distance of our home and learned to appreciate and care for all the tropical vegetation so different from New England.

Lillian Race Williams with wives who lived in her home. *Photo courtesy of the Boca Raton Historical Society.*

THE ROLE OF CIVILIANS AND WORKING WOMEN

Civilians were vital to the overall operation of the base and served in practically every office and department. Averaging 1,200 and peaking at 1,500, most worked in the academic, sub-depot, quartermaster and post engineering departments.

Because of the lack of sufficient civilian workers in Boca Raton, personnel were drawn from Delray Beach, West Palm Beach, Deerfield Beach, Pompano Beach and Fort Lauderdale. Four civilians crucial to the morale of the troops were post office supervisor Ralph Ralston as well as postal clerks Sue Archer and Richard and Eldra Durham, a married couple, sent up from the Fort Lauderdale post office.

The Durhams, who served on the base from 1943 until it closed in 1947, worked between ten- and eleven-hour days. They sold stamps and money orders, and delivered to troops hundreds of thousands of packages and letters, bearing both good news and bad.

Living in Fort Lauderdale, the Durhams commuted eighteen miles a day up U.S. Route 1 or Dixie Highway, a half-hour drive. To shorten their commute, they rented a three-bedroom house on Palmetto Park Road in 1944. Their rent was $37 a month, which included water and electricity costs as well as an $8 monthly charge for furniture rental.

As with all installations, occasionally there were alarming incidents. Eldra recalled one of those times:

Singing Pines, Lillian Race Williams's beloved home, is now the Boca Raton Children's Museum. *Photo courtesy of the Boca Raton Historical Society.*

Servicemen found accommodations for their families wherever they could. Waino and Doris Ray show off their new addition, Sally, in front of their rented beach home "Poinciana." It was owned by Boca Raton resident Clara Riggs. *Photo courtesy of Waino and Doris Ray.*

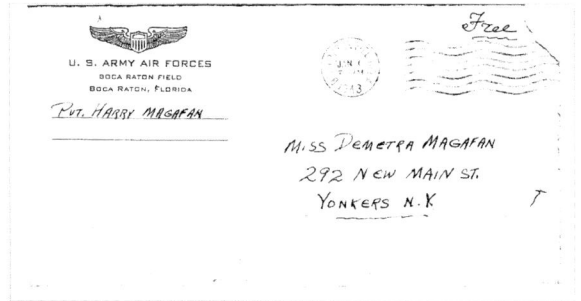

Left: Every base in the U.S. had its own stationary. This is the official stationary used by Army Air Force officers. The logo was embossed in gold-colored ink. *Photo courtesy of the Boca Raton Historical Society.*

Right: Harry Magafan sends a letter to his sister in Yonkers, New York. The postmark reads: "Boca Raton Field, Fla., January 7, 1943, 7 AM." *Photo courtesy of Harry Magafan.*

A letterhead penned by one of the soldiers depicts a bomber using radar to locate a German U-boat. *Photo courtesy of the Boca Raton Historical Society.*

We had a peeping Tom. Three of us girls went to the bathroom and all of a sudden Ann Bennett looked up and there was a man laying on the platform above watching us. She said, "What the ——— are you doing up there?" He jumped down, went through the window and up the hill to a barracks with all three of us after him. When we got to the barracks, he went on through with nobody stopping him and went on to the next one where they stopped him. They had a trial later on, but we wouldn't go to the trial. We were a little afraid of what he might do. They sent him off the base.

When the base closed in 1947, all postal equipment was returned to Fort Lauderdale. The combination safe, borrowed from the Boca Raton post office, was also returned, and the Durhams moved back to their Fort Lauderdale home. In 1950, Richard Durham received a curious call from the Boca Raton post office, asking if he could recall the combination to the base safe. The post office staff said the safe had not been opened since the base closed. Thankfully, he remembered the combination—10-15-19-42—the date the base opened.

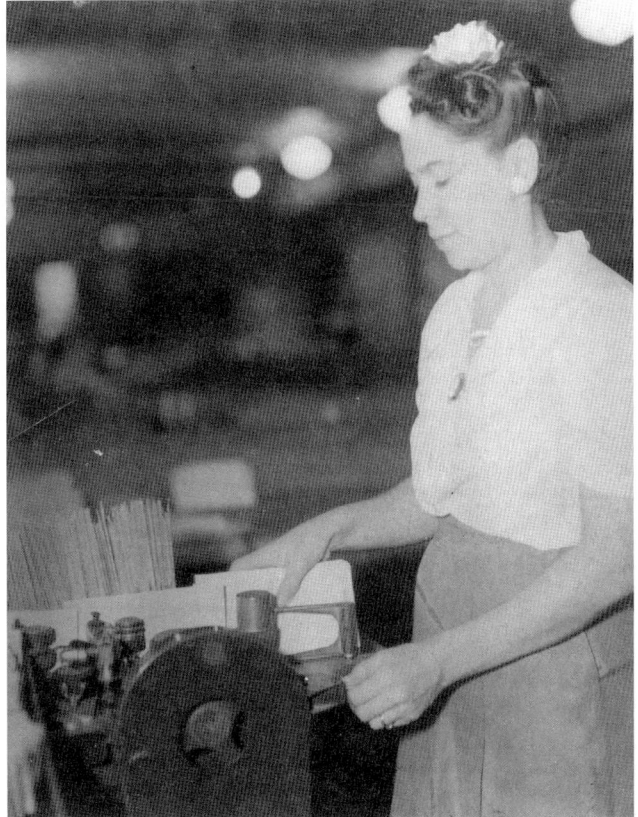

Eldra Durham, a postal clerk at the Boca Raton Army Air Field, runs mail through the postage meter. *Photo courtesy of Eldra Durham.*

WOMEN AND THE WAR EFFORT

Women played a vital role during the war as they entered the workforce in record numbers, filling jobs vacated by men who joined the armed services. "Rosie the Riveter" became a national icon and encouraged women to participate in the war effort, assuring them that working in the defense industry would not compromise their traditional feminine roles.

Local women worked and volunteered their services in the liveliest joint on the base—the Service Club. Opened on December 19, 1942, it had elaborate facilities designed for "diverting the weary minds of service men and women from the trials and tribulations of military life," according *Boca Raton Radar School*.

The Service Club hosted its first formal function in 1942, a Christmas dance complete with a large tree. An article in the *Transmitter* stated: "Complete facilities of the Club provided for dancing, movies, stage shows, reading, writing, voice recording, billiards, various artistic hobbies, ping-pong, archery, horseshoe pitching, cards, chess and other table games. Musical facilities include three pianos, a music room with large libraries of popular and symphonic recordings and radio."

When the cafeteria opened in the Service Club on January 5, 1943, base personnel got some relief from the mess halls with economically priced meals, sandwiches, snacks and a soda fountain. The library was stocked with more than 4,500 books and 100 popular and technical magazines. Civilian women served as the Service Club director, librarians, and recreational and cafeteria hostesses.

The Service Club, the liveliest joint on the base, served as a gathering place for enlisted men and women. *Photo courtesy of the Boca Raton Historical Society.*

AMATEUR BOXING A WEEKLY DIVERSION

Along with frequenting the Service Club, the men watched—and some participated in—amateur boxing. Held on Monday nights, the bouts were divided into weight classes. Fighters trained in the gym and fought in an indoor arena and outdoor stadium.

One of the boxers was Charlie Romero. He came to Boca Raton from St. Petersburg, Florida where he served with such notables as Clark Gable, Jimmy Stewart and Glenn Miller. He served at the Boca Raton Army Air Field from January 1943 to March 1944 guarding bombsights, airplanes and radar equipment.

Romero recalled his military boxing career:

> Of all the sports that I loved, I decided on boxing. It would be fun to learn and entertain the troops. I trained hard with my trainer, Mike Romano. I worked myself up to the heavyweight championship fight against the current base titleholder, which I happened to beat in forty-eight seconds in the first round. From then on, they started call me Charlie "John L." Romero, because of my stance like John L. Sullivan. My left jab was my best asset, along with lots of power in my punch. Everyone on the base knew me because of that fight.
>
> At that time, my trainer wanted me to fight at Madison Square Garden after the war. The GIs were already matching me with Billy Conn, who was in line to fight

Charlie Romero served as a base guard. *Photo courtesy of Charlie Romero.*

Amateur boxing took place on Monday nights. Charlie Romero fought in the heavyweight division. *Photo courtesy of Charlie Romero.*

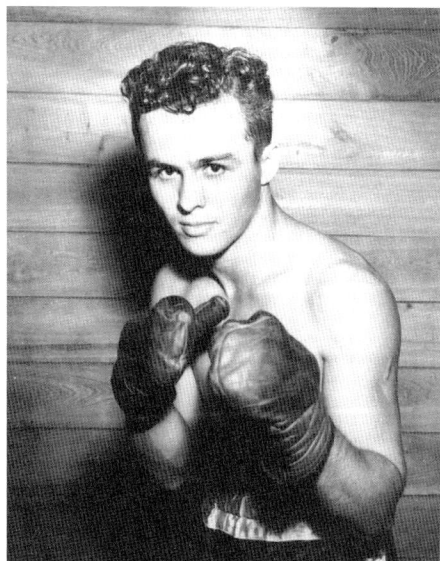

Ed Malavarca fought in the featherweight class. He never lost a bout. *Photo courtesy of Ed Malavarca.*

the heavyweight champ of the world, Joe Louis. But after seeing Chester McGursky, the champ of the Armed Services that I beat, I regretted hurting another person that badly. He was in the hospital for two and a half weeks with broken ribs and the like. My heart and conscience wouldn't let me go on boxing. I quit.

After serving in England, Romero returned to the United States aboard the *Queen Mary*, along with fifteen thousand troops and a lone congressman. The trip made newspaper headlines. Losing a little weight, Romero resumed his boxing career later in Georgia and won the light heavyweight championship.

Ed Malavarca fought in the featherweight class. Of his fifteen matches, he remained undefeated and became the self-described "champ" in his weight class.

"I participated in boxing, which afforded me a chance to get out of there on weekends. I and another couple of chosen GIs would rent a car and take off for the hinterlands," Malavarca said.

BASE FOOTBALL TEAM PLAYS MIAMI HURRICANES

To encourage participation in athletics, the base formed several football teams. Troops scrimmaged with each other and played teams from other military bases. Twice in 1943, games were scheduled against the Hurricanes from the University of Miami.

The University of Miami Athletic Office notes that in the same year the Hurricanes played other military installations with the following results: Jacksonville Navy Auxiliary Air Training Center (UM 6–0), Camp Gordon (UM 52–6), Charleston Coast Guard (UM 13–6), Jacksonville Navy Auxiliary Air Training Center (Jax 20–0), and Fort Benning (UM 21–7).

BOCA RATON AND THE AIR FIELD: A SYMBIOTIC RELATIONSHIP

Not only did local businesses profit from providing goods and services to the Boca Raton Army Air Field, the community profited as the base made recreational and medical facilities available to local civilians. Residents mingled with service personnel at the local bars and stores, and met at various volunteer organizations. Local women and army wives became spotters at towers along the beach where they searched the coast for enemy submarines, spies and aircraft. They also planted victory gardens, bought and sold war bonds, and collected metal toothpaste tubes and tin cans for recycling.

The *Transmitter*, reports in a May 14, 1943 article: "Approximately 250,000 square feet of ground will be marked off into numbered plots, which will be cultivated with a wide variety of Florida produce by 300 officers, nurses and enlisted men of Boca Raton field.... All gardening materials will be purchased collectively and the garden space will be cleared and planted in collaboration with adjoining civilian garden clubs. Produce

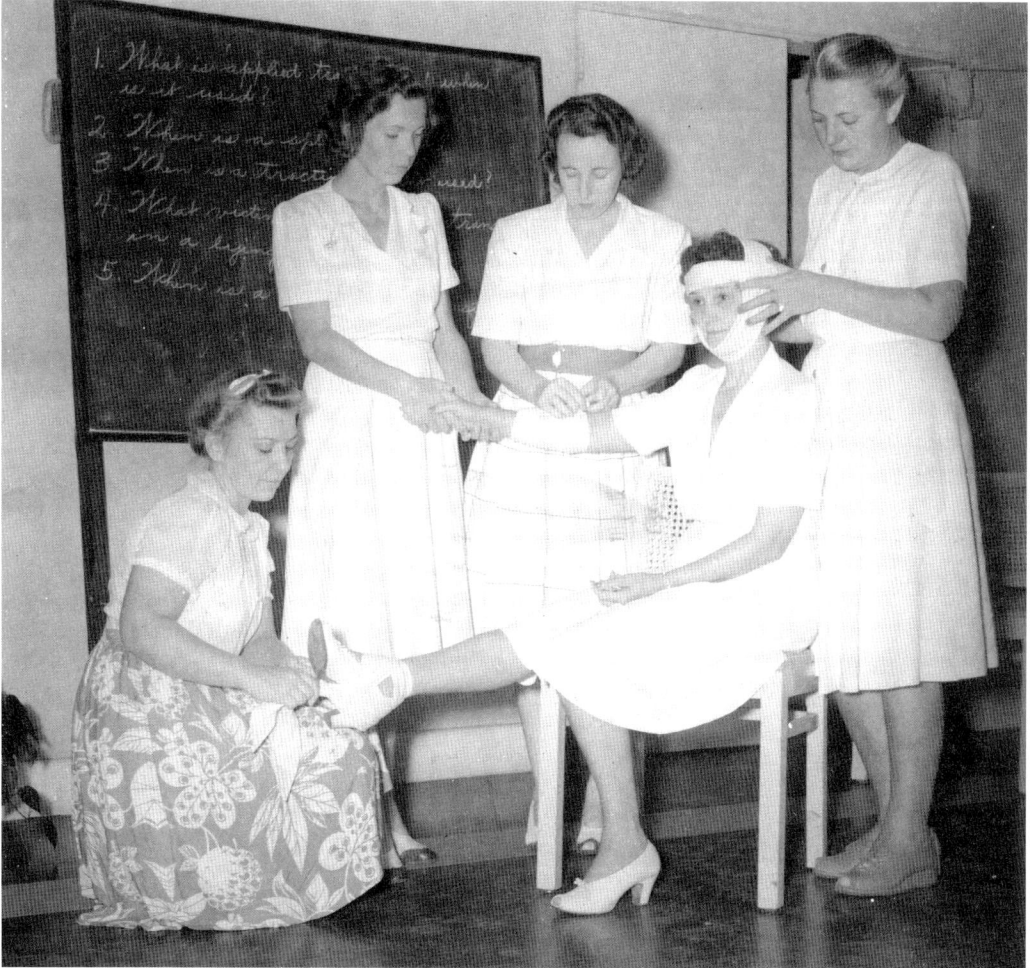

Civilians practice first-aid techniques taught by the Red Cross. *Photo courtesy of the Boca Raton Historical Society.*

harvested from individual plots will belong to the gardener himself and can be used any way except commercially."

As was the case across the nation, many Boca Raton women contributed to the war effort by becoming volunteers. The Red Cross trained many women in first aid, home nursing and emergency preparedness. The Boca Raton Town Hall hosted knitting sessions, and Mrs. Harold Butts and Mrs. T.M. Giles directed a Red Cross sewing room for national-defense volunteer work in the Mitchell Arcade, a multi-purpose two-story building with store fronts downstairs and apartments above Even young girls got involved, joining the Junior Red Cross that sponsored dinners and made stuffed toys to send to British hospitals.

Initially composed of 53 musicians, the 503rd AAF Band grew to 170 members. Smaller jazz bands, string quartets and dance bands played at civilian and military venues throughout south Florida. *Photo courtesy of the Boca Raton Historical Society.*

BOCA RATON FIELD BAND

Civilians and troops weathered the storm of war by enjoying the base's most beloved entertainment group: the 503rd AAF Band As band director, Captain Maurice Shorago—a service pilot authorized to fly utility aircraft—flew around the country, recruiting members from celebrity bands. Manny Chavez recalled: "He would tell the band member that if he didn't volunteer to come to Boca Raton, he would have them drafted into the infantry."

Musicians from Jimmy Dorsey, Abe Lyman, Tommy Dorsey, Glenn Miller, Bob Crosby, Artie Shaw, Benny Goodman and Paul Whitman dance bands and orchestras, along with other good amateur musicians, made up the band. Initially, it involved 53 musicians who played at official ceremonies and engagements. As the band grew to 170 members, musicians grouped into smaller units—a symphony orchestra, several string quartets, woodwind octets as well as dance and jazz bands—that played throughout South Florida.

Should the troops think serving one's country in wartime as a member of the band was a "cush" job, they could refer to the *Transmitter* that published this rehearsal and performance schedule:

Members of the 503rd AAF Band horn section. *Photo courtesy of the Boca Raton Historical Society.*

Day in the Life of Army Musicians
Reveille – you know when
Rehearsals from 8:30 to 11:00 a.m. daily
Chow
Dress and fall out for Formal Guard Mount at 2:45 daily
Fifteen minutes of playing for F.G.M.
Hike to one of the official drill fields for daily Retreat and Review (at which time the colonels and the majors and the captains and the lieutenants watch with eagle eyes for the slightest defaults.)
After Retreat, Chow.

Through for the day?
Hardly!

Music From Morn 'Til Night
Every Monday night there is a full concert in the Pinetree Bandstand. All 173 pieces, stacked in tiers on the stage. During the rest of the week, the band is divided into units for playing jive at numerous squadron affairs. The two main dance bands are "The Bombardiers" and "The Swing Pilots."

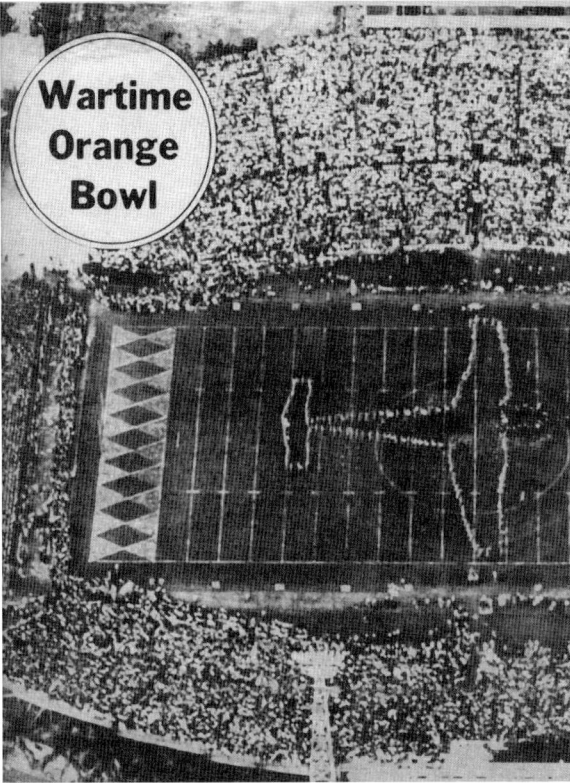

The 503rd AAF entertains the crowd during halftime at the 1943 Orange Bowl game. *Photo courtesy of the Boca Raton Historical Society.*

On several occasions, the group teamed up with other bands from South Florida bases to perform in parades or special athletic functions. Such was the case on January 1, 1943, when a three-hundred-piece band—formed from the Boca Raton 503rd AAF Band, the Officer Candidate School Band of Miami Beach and the famous Navy Band—entertained spectators at the Orange Bowl football game between Boston College and Alabama. The bands took the form of an airplane that moved across the field.

CLOSING OUT 1943

The town of Boca Raton and the army air field closed out 1943 on a positive note. The base brimmed with troops, commerce in Boca Raton and the surrounding area thrived, local residents enjoyed base services and the number of civilian jobs rose steadily. Additionally, aircraft arrived at the field, increasing the quality of pilot training; progress at the Rad Lab allowed new equipment to arrive to improve radar technical training.

Everything seemed to be coming up roses for Boca Raton and the air field. But, as both would see, the arrival of so many troops to this unfamiliar and unprepared area was about to cause considerable strain.

1944 Time Capsule

The *Batman & Robin* comic strip premieres in newspapers • Congress charters the Central Intelligence Agency • The first TV musical comedy, *The Boys from Boise*, debuts • The John Hopkins Hospital performs the first open-heart surgery

World News

• D-Day begins when allied troops make landings at Normandy, France, which leads to the recapture of Paris.
• The first V-1 rocket bomb hits London, England.
• General Douglas MacArthur returns to the Philippines
• German officers make an unsuccessful attempt to assassinate Adolph Hitler.
• French tanks lead Allies into Paris.

National News

• Franklin Roosevelt is elected to his fourth term as president.
• Harry S. Truman is elected vice president on the same Democratic ticket.
• Big Bend National Park is established.
• The GI Bill of Rights is passed, giving World War II vets benefits, most notably housing subsidies and funding for college education.
• Meat rationing ends in the United States.
• *Seventeen* magazine makes its debut.
• Chiquita bananas are introduced.

Sports

• World Series champion—St. Louis Cardinals • U.S. Open golf champion—Was not held in 1944 • Pro football champion—Green Bay Packers • Indianapolis 500 winner—Was not held in 1944 • Stanley Cup winner—Montreal Canadians • NCAA basketball champion—Utah • College football champion—Army • Heisman Trophy winner—Leslie Horvath from Ohio State

Music

• "Don't Sweetheart Me"—Lawrence Welk
• "G.I. Jive"—Louis Jordan
• "I Dream of You"—Andy Russell
• "I Love You"—Bing Crosby
• "I'll Get By"—Harry James
• "I'll Walk Alone"—Dinah Shore
• "It's Love-Love-Love"—Guy Lombardo
• "Mairzy Doats"—Merry Macs
• "My Heart Tells Me"—Glen Gray
• "Shoo Shoo Baby"—The Andrews Sisters

Movies

• *Going My Way*—Academy Award winner
• *Jane Eyre*
• *Double Indemnity*
• *To Have and Have Not*
• *Gaslight*
• *Meet Me in St. Louis*

Cost of Living

New house	$3,475.00	
Average income	$2,378.00 per year	
New car	$975.00	
Average rent	$50.00 per month	
Tuition at Harvard University	$ 420.00 per year	
Movie ticket	$.40 each	
Gasoline	$.15 per gallon	
U.S. postage stamp	$.03 each	

Food

Granulated sugar	$.75 for 10 pounds
Vitamin D milk	$.62 per gallon
Ground coffee	$.48 per pound
Eggs	$.21 per dozen
Fresh baked bread	$.10 per loaf

1944

Full House

In 1943, the Boca Raton Army Air Field served an average of eleven thousand troops; however, that figure jumped to more than fifteen thousand in 1944, as the number of men and women needing to be trained increased dramatically.

At the Boca Raton Club, an additional one thousand officers were expected to enroll in the Radar Officer (Air) Training Course, increasing the number of personnel in the area six-fold. Unable to provide housing for the additional personnel inside the club, plans were made to include a tent camp on the golf course and move small portable buildings up from Miami Beach.

On the Boca Raton Army Air Field, similar problems arose, as reported in *Boca Raton Radar School*: "During the months of September and October, the student flow increased approximately 50%, being fairly regular during September at about 300 men per week and increasing during October to a maximum of 500 per week. This has necessitated opening of barracks which had been standing idle in the north area for approximately one year."

With construction on base mostly completed and the base now the temporary home to a substantial number of troops, routine daily operations increased as well. This required more civilians to cope with the tremendous number of troops transitioning on and off of the base.

But civilian workers were hard to come by. Boca Raton couldn't supply all that were necessary; and with gasoline rationing, travel from West Palm Beach or Fort Lauderdale proved difficult. There were never enough workers to perform necessary tasks, a problem reported to headquarters on a regular basis.

Obtaining off-base housing remained a problem for military personnel and their dependents; blacks and Jews found it especially difficult as many landlords would not

Planes line the field at the Boca Raton Army Air Field. *Photo courtesy of the Boca Raton Historical Society.*

rent to them. Price gouging became another problem. Rental rates got so out of hand that Army Air Force Headquarters in Washington, D.C., requested information on rent conditions in the vicinity of all its installations.

Boca Raton Radar School gives an idea of just how bad the situation became:

> *The Commanding Officer of the BRAAF replied by giving a detailed account of the local situation. He also recommended that an effective rent control, designed to meet the problems peculiar to this area, should be promptly established. Rental control by the cities in cooperation with the Army and Navy was not successful. Several*

cases were cited to prove the serious nature of the situation. On 19 November, a key civilian employee who had rented an apartment for $40.00 per month on a year round basis in Delray Beach was asked to leave in order to make room for tourists. She was offered an efficiency apartment at the rate of $450 a month. As a result, she was forced to request a release from her position from the base. Similarly, a Second Lieutenant was asked to pay $500.00 per month or give up his apartment. In both cases the occupants had been welcomed during the tourist season of 1942-1943.

FLYING HIGH

When the base first opened, only 10 dilapidated English Hudson patrol bombers were sent to the base. But by August 1944 the number of aircraft permanently stationed at the base reached 117. These aircraft were:

Type of Plane	Number of planes at Boca Raton	Number of Engines	Name of plane
AT-6	1	Single	"Texan" advanced trainer (most popular aircraft to train pilots)
AT-10	1	Two	Advanced trainer
C-78	5	Two	"Bobcat" cargo plane
AT-11	12	Two	"Kansan" advanced trainer
B-34	20	Two	"Lexington" bomber
B-17	61	Four	"Flying Fortress" bomber
B-24	8	Four	"Liberator" bomber (bigger and faster than B-17)
Other	9	Four	

Other aircraft that came and went from the base included the B-25 "Mitchell," B-18 "Bolo," P-51 "Mustang," B-26 "Marauder," P-47 "Thunderbolt," C-47 "Gooney Bird" and the B-29 "Superfortress."

Radar was installed on the aircraft, and troops learned to use the newly installed equipment by practicing bombing runs over the Atlantic. To ensure aircraft safety, airplane mechanics ran nose-to-tail inspections while electronics mechanics tested communications systems.

Charles Bender arrived at the Boca Raton Army Air Field, having previously "washed out" of cadet school in Miami Beach. He was subsequently sent to Radio School at Truax Army Air Base in Madison, Wisconsin, where he became an instructor. He told about his time at Boca Raton:

Mechanics change a propeller on a B-17 "Flying Fortress," one of the most famous airplanes ever built. The aircraft served in every World War II combat zone. but is best known for daylight strategic bombing of German industrial targets. *Photo courtesy of the Boca Raton Historical Society.*

After graduating from radar school, I was made aircrew and started flying on B-25s and B-17s in order to check out radio and radar equipment. I flew on routine submarine patrols off the coast of Florida in B-17s, probably the sturdiest bomber made up to that time. It was at this time I saw my first B-29. This was an astounding airplane and dwarfed the B-17s parked nearby. We were told it was a new long-range heavy bomber and was here to be outfitted with radar-navigation equipment and a radar bombsight, the ME2-13, that we called the "Mickey" bombsight.

Little did we know that a plane like this would drop the bombs that ended the war in the Pacific. I flew on B-29s a few times to check the radar system but did not like having to inch my way through the tunnel that connected the plane to the forward section where the equipment and flight deck were located. We'd heard stories of crewmembers being propelled through the tunnel like a projectile due to sudden changes in air pressure.

Louis Delgado wasn't a pilot, but he yearned to go on training missions. "I remember also wanting to fly so I would get approval to go on night training flights to as far as

This B-24 "Liberator" sports its pilot's sentiments in nose art, common during World War II. More B-24s were built than any other wartime aircraft. With its great range, it performed anti-submarine work in the Atlantic and heavy-bomber support in the Pacific. *Photo courtesy of the Boca Raton Historical Society.*

A B-25 "Mitchell" sits on the tarmac at the Boca Raton Army Air Field. While the airplane was originally intended for level bombing from medium altitudes, it was used extensively in the Pacific for bombing Japanese airfields from treetop level as well as for strafing and skip bombing of enemy shipping. *Photo courtesy of the Boca Raton Historical Society.*

This Piper aircraft was used in training and for utility runs. *Photo courtesy of Lt. Col. Manuel Chavez USAF (Ret.).*

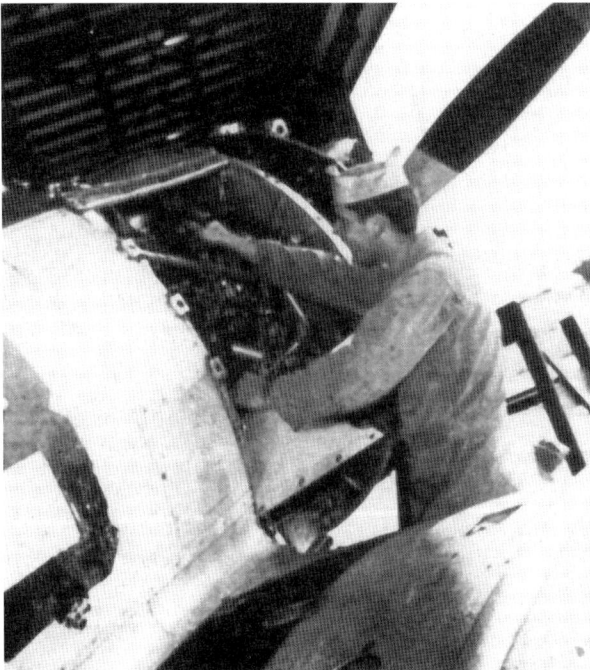

Flight-line personnel perform tasks to support those in the skies. *Photo courtesy of the Boca Raton Historical Society.*

Charles Bender in the co-pilot seat of a B-25. *Photo courtesy of Charles Bender.*

Cuba and back. My training for the flights consisted of being shown how to strap on a parachute and how to deploy it in the event of a need to bail out. The B-25's that I flew in were so noisy that if I screamed at the top of my lungs, I wouldn't be able to hear myself," he said.

MAY DEADLY FOR PILOTS

May 1944 proved to be a deadly month for the Boca Raton Army Air Field. On May 12, nine men were killed when their light bomber crashed and burned while taking off on a routine training flight. Eight of the men died instantly, and one succumbed later in the post hospital. Those who died were First Lieutenant William H. Carson, the pilot; First Lieutenant Thomas A. Lamont; First Lieutenant John J. Lominac; First Lieutenant Benjamin P. Sibley; Private First Class Norman R. Stiner; Sergeant John J. Sasicko; Private First Class Robert E. Locke; and Staff Sergeant Frank L. Bursaw.

On May 30, Colonel Edward S. Allee, thirty-six, former deputy commander for training and operations at Boca Raton Army Air Field, and Lieutenant Colonel Gilmer T. Smith, director of the ground school, died when their C-78 went down on a routine instrument-proficiency flight seventeen miles northwest of West Palm Beach, near Jupiter, Florida.

The aerial view of the airfield looking west. *Photo courtesy of the Boca Raton Historical Society.*

Veteran pilot Manny Chavez was an eyewitness to several crashes. One involved training on B-34s. According to Chavez, the aircraft was difficult to fly because it was "heavy and bulky, with short wings and powerful radial engines. If an engine stopped on takeoff, the pilots had to reduce power on the good engine to keep from rolling over."

"We lost two of these planes on takeoff because of this problem, one of which was flown by one of my classmates. I was right behind him, ready for takeoff, when he crashed and died with all the crew a quarter mile from the end of the runway. I was cleared to take off and had to pass right over the top of the wreckage," recalled Chavez.

This Piper was one of two that crashed over the Florida Everglades when the young Boca Raton Army Air Field pilots practiced dogfighting among the buzzards. *Photo courtesy of Lt. Col. Manuel Chavez USAF (Ret.).*

Chavez also witnessed a head-on collision between a B-24 and B-17 one morning before daybreak.

> *The B-24 pilot was an experienced combat pilot who had just returned from Europe. The B-17 was taxiing up the northwest segment for takeoff to the east, and the B-24 was headed up the southwest runway, also for the east takeoff. The uncompromising rule is that when you reach the takeoff position, you stop and read the final checklist out loud so the other can hear it. Then you check your compass heading to confirm that you are on the correct runway.*
>
> *The B-24 hotshot combat pilot made a running turn to the takeoff runway, but instead of getting on the east runway, he got on the southeast runway and locked wings with the B-17 about one thousand feet down the runway. I was on the ramp in front of operations, and it was a spectacular crash, shearing the cockpit off the B-24 and breaking the B-17 in two. I saw one of the pilots run a one-hundred-yard dash away from the crash with a parachute strapped to his back.*

Daily Routine

Veterans and civilians alike recall Hollywood wartime movies depicting troops marching to songs. Bill Robison explained the foundation for this:

> With regard to marching songs, the Army, during *WWII*, was committed to a singing army. It was very much believed by psychologists that this was essential to maintaining a high level of morale among the troops. Toward this objective, new recruits were issued songbooks during basic training so that there would be no excuse for a soldier not to join the singing. Drill sergeants would pass back and forth along a marching column to make sure each and every soldier was singing. Thus, pretty much from reveille to taps, music was in the air at BRAAF as trainees marched to and from radar classes and training facilities such as physical training, rifle range practice, and enemy aircraft identification, etc.
>
> One singing quirk which might have been unique to BRAAF was that whenever a body of marching men passed by a column of marching WAACs (Women Army Corps), the men would take off their helmets (actually helmet liners), wave them at the ladies and sing, over and over—
>
> "For we're the joy boys of radio*,
> Hello, hello, hello, hello."

*Radio is the basic and underlying technology upon which radar is based.

Daily routine for troops at the base always began with PT (physical training). No matter the season or weather, troops marched in formation to the PT-designated areas and engaged in a rigorous workout.

Sometimes getting to the fields in the heat of the summer proved difficult. Philip Kantor recalled: "I seem to remember marching on what appeared were tar-based roads, which were difficult to march on because the soles of the army boots would stick to the hot tar surface."

With the heat and constant physical exertion came heavy perspiration. Ed Wallish confessed that even showers didn't always solve the problem: "No matter how often you showered, it didn't take long to build up another sweat. It was impossible to change clothes more than once a day, and everyone smelled from stale sweat."

Though troops who arrived during the winter months didn't have to contend with the heat of the summer, they did encounter cold spells. Wallish recalled: "It was cool during the winter months; nevertheless, we had to wear khakis and could not wear our field jackets. Sometimes I wore a sweater or extra T-shirt to keep warm, but it was shorts and T-shirts for daily calisthenics."

Troops scale a wall during physical training. *Photo courtesy of the Boca Raton Historical Society.*

Traversing a hazard using ropes was all part of daily PT. *Photo courtesy of the Boca Raton Historical Society.*

Paul Metro (right) and buddies on bivouac training. *Photo courtesy of Paul Metro.*

Mornings seemed to be the most difficult time for the troops, according to Paul Metro:

> *The first weeks were difficult to get used to. We were awakened at 4:00 a.m., then we washed, shaved and dressed, and proceeded to the assembly area on the parking lot by the squadron supply building. We were so tired in the morning that we would lie down on the blacktop to catch a last few winks of sleep.*
>
> *After assembling our tired bodies out in the parking lot, we were then marched double time onto the road for PT. Everyone tried to get to the rear of the formation and as far away from the instructor as you could. It still being dark, men in the back could not be seen; this way the unlucky ones in the front of the formation were sweating and struggling through side-straddle hops and push-ups while those clever soldiers in the rear snickered and only pretended to be engaged in the rigors of physical exercises. After PT, we marched on to breakfast.*

The troops experienced additional training both on and off the base. Metro recalls bivouac training: "Additional training included a three-day bivouac in the backwoods of the Everglades. We hiked with full backpacks, carrying an empty carbine along dusty roads and trails. We slept on the ground in pup tents and took turns at guard duty during the night. One night, my tentmate scared me when he woke me to take my turn. He gently and quietly touched my foot, which I thought was a snake or other animal. He apologized for my alarm. Our meals were K and C rations."

Dining in the Mess Hall

After PT, troops marched to the mess halls for breakfast. "They continued marching wherever they went, singing songs such as, 'I've Been Working on the Railroad,' 'My Gal Sal,' 'I've Got Six Pence' and comical songs denigrating Army life such as 'Gee Mom, I Wanna Go Home,'" Wallish remembered.

A railroad spur off the Florida East Coast tracks allowed freight cars to deliver tons of cold meats and foodstuffs to refrigerated warehouses. Vegetables, breads and other supplies arrived by the truckload into central receiving, where they were divided and routed to the various mess halls. Cooks and kitchen staff peeled potatoes, cut up chickens and roasted sides of beef to feed thousands of hungry soldiers.

Wallish remembered the buildings that served troops twenty-four hours a day: "The mess halls were large plus (+) shaped wooden buildings covered in green tarpaper. The vertical members were two dining halls, the horizontal members

Crates of fresh bread arrive at the loading docks from Holsum Bakers. *Photo courtesy of the Boca Raton Historical Society.*

Celery, onions and other fresh vegetables arrive for distribution to the eleven mess halls. *Photo courtesy of the Boca Raton Historical Society.*

A hungry soldier digs in. *Photo courtesy of the Boca Raton Historical Society.*

contained kitchens, storerooms, offices, toilets and a coal fire boiler room that supplied hot water for the kitchen and dishwasher machines (unexplainably called 'clippers,' thus if you were a KP [kitchen patrol] washing trays or dishes you were said to be 'flying the China Clipper')."

Metro recalled having to wait for the mess halls to open and laments the quality of the food.

After PT, we lined up at the portals of Mess Hall 3, waiting for it to open up and serve breakfast. Some mornings the mess hall didn't open up on time, and the hungry troops would begin to yell impatiently. Whenever this happened, there would be barracks rumors that there was trouble with the KPs, who were all black.

I thought the food was horrible. I ate only the boxed cereal, toast with butter or jam, fruit and milk for breakfast—never the scrambled eggs or coffee, which usually had a greenish tinge, probably due to saltpeter being added. I didn't eat the SOS (chipped beef on toast), although many guys did. It just didn't look appetizing to me. I was very choosey with other meals, and I paid for many at the Service Club—it was worth it.

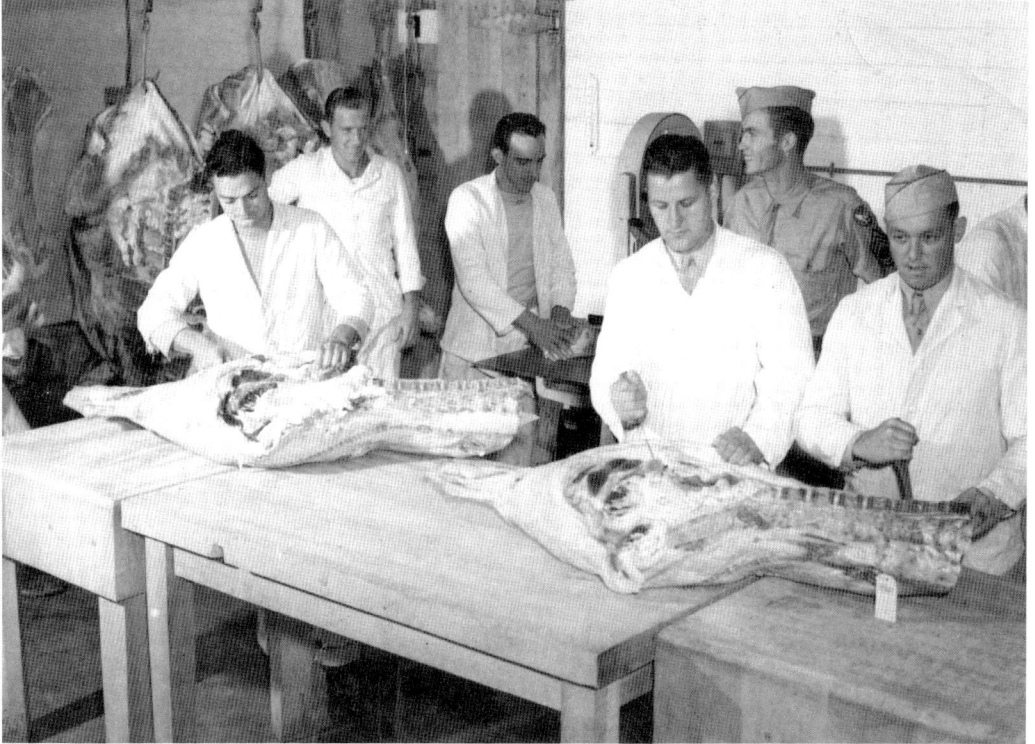

Sides of beef arrive at the refrigerated warehouse where the meat is cut up and dispersed to the mess halls. *Photo courtesy of the Boca Raton Historical Society.*

FOOD WASTE

One of the biggest concerns on the base was what to do with food waste from the mess halls. With more than 447,703 meals and 616,126 pounds of food served in June 1944, commanding officers closely watched "plate waste" and applauded any decrease compared to the previous month.

Even carcass fat from beef was monitored. A shortage indicated a more lean supply of beef; however, it also meant the kitchen had a shorter supply of rendered fat for cooking.

KP AND OTHER DUTIES

Radar training wasn't the only job troops had on base. While awaiting the beginning of radar classes, newly assigned troops were considered "casuals" and pulled what was known as "casual duty." This included such mundane tasks as KP (kitchen patrol), trash

Cooks bake rolls by the pan full to feed hungry soldiers. *Photo courtesy of the Boca Raton Historical Society.*

pickup, supply delivery, garbage detail or other general duties. Duty sergeants were always looking for men to fulfill never-ending needs for distasteful work.

Ed Wallish recalled time spent on casual duty:

> *Policing* [cleaning up] *the squadron area was a task that seemed to come around once a week or more. It would start with the men of the squadron being assembled in the parking lot by the squadron supply building. The Squadron TO2 Commander, Captain William F. Burford, would then talk to us and describe in excruciating and infinite detail how he wanted every cigarette butt, every candy bar wrapper, and every cigarette package picked up and disposed of. We would then form a single row along the road and proceed west through the squadron area, picking up stuff as we went.*
>
> *The squadron area was on sandy ground and the remains of every piece of trash that had been dropped since October 1942 was buried in the sand. As we walked along picking up trash, our feet would disinter refuse from the sand behind us. As we proceeded, Captain Burford inspected our efforts with great attention to detail.*

After our first policing run, Captain Burford reassembled us on the parking lot and chastised us for not doing the job right. We would have to do it again. Once again we went through the area excavating more 1942 trash. After policing the area for a second or third time, we sweated in the hot sun and listened to Captain Burford deliver yet another boring orientation lecture.

Paul Metro remembered:

Every morning after we finished school training, we assembled for work details. This included miscellaneous manual labor such as repairing boardwalks, raking sand, KP and moving coal. A coal-fired boiler in a shed attached to the latrine provided hot water. We had to move coal dumped in the road into the coal bin.

KP was from 0200 to 2100 hours—it was torture. Those on KP duty tied a towel at the foot of their bunk, so that the CQ (charge of quarters) knew whom to awaken. KP included washing down the tables and mopping the floor after every meal, washing pots and pans, and cleaning service trays. We separated the garbage into "edible" and "nonedible." The edible was sold to farmers for hog feed.

KP included cleaning the large vats used to make soups and stews. *Photo courtesy of the Boca Raton Historical Society.*

Scheduled for KP, Louis Delgado found an unorthodox way to avoid it:

There was the time one payday when I couldn't leave the base because I was scheduled for KP early the next morning. So I ended up getting drunk on base with a few drinks some of my buddies gave me. When I arrived at the mess hall, I was so dizzy that the mess sergeant told me to lie down next to the mess hall entrance until I felt better. I fell asleep next to the doorway and when I woke up, breakfast was being served. I then proceeded to get in line. For some reason, no one noticed me there or just decided to ignore me.

Troops caught disobeying their superiors were reprimanded and required to perform tasks designed to instill discipline and obedience. Ned Renner described such a time:

One of the students was caught by a corporal throwing a cigarette butt on the ground. What a mistake! It was well into the night by the time he dug a hole measuring six feet wide by six feet deep in which he placed that cigarette butt. He was then ordered to fill the hole back in to bury the cigarette butt.

On July 3, 1946, I reached the age of 17, and a few of us decided to go to Fort Lauderdale. It was not that we drank a lot, we merely wanted to get away from the base and enjoy ourselves. Bed check was at 2400 hours, and we were not in our beds for bed check. This was real trouble from the gitgo. The following morning the 1st Sergeant reprimanded us and prescribed a close order drill for eight hours on the following Saturday morning. It was the most grueling period of time I have ever spent on a drill field.

The drill instructors changed every hour, and they were fresh. Ambulances removed soldiers from the drill field and transported them to the hospital for medical attention. We got a 15-minute break each hour where we would smoke and listen to the drill Sgt. read [the] Articles of War.

I was a tough little brat who grew up in the slums of Hagerstown, NJ and at the end of the day I got my second wind. After evening chow I walked into the Orderly Room and asked the 1st Sergeant if I could get a pass to leave post. He said to me, "If you lived through this day and still want to go on pass, you've got it." I was punctual for bed check after that.

Following that incident, the Post Commander got into a lot of trouble because he had subjected several hundred soldiers to mass punishment, an act forbidden by the Articles of War.

BARRACK INSPECTIONS

Barrack inspections occurred once a week, usually on a Friday or Saturday morning. The night before, the men would spend their time cleaning up, polishing shoes, making beds and neatly arranging their footlockers. Those who passed inspection earned a weekend pass. Those who didn't were assigned casual duty.

Ed Wallish remembered the precision with which the inspections were made: "We wore clean heavily starched khaki uniforms, were freshly shaved and had new haircuts. There were shelves and clothes racks built into the walls, and our clothing had to be fully buttoned and hung in a prescribed order. We awaited the inspectors alongside our bunks and came to attention as Captain Burford entered the barracks. Then, he and Lt. Kuhn and Master Sergeant Westmoreland would make the rounds looking for infractions. After inspection, we were usually free for the weekends unless you were unlucky and detained to pull KP."

Paul Metro remembered getting demerits for not having his hair cut short enough: "It was so common a demerit, we joked that the officers were getting a payoff by the barbers. After getting a haircut and reporting to the orderly room, the demerit was removed, and the weekend pass released."

Chute Landing School Added

In June 1944, the Boca Raton Army Air Field added a Parachute Landing School for the training of all flying personnel. The twelve-hour course was introduced to cut down on the number of air casualties that resulted from a lack of knowledge on proper parachute jumping methods, and instruction became an integral part of the physical fitness program.

The course was divided into five phases. It began with tumbling fundamentals and moved on to cover preliminary jumping from platforms that ranged in height from four feet to sixteen feet. There, the men learned proper landing techniques.

In phase three, the men strapped on a parachute and were suspended from a harness. In this position, they learned the correct fitting of the chute as well as how to swing, turn, make water landings and avoid high-tension wires.

The fourth and most exciting phase used the landing trainer, a harness attached to a trolley on an eighty-foot-long cable. At its apex, the cable was thirty feet from the ground; it tapered down to nine feet.

The fifth phase used a wind machine to teach personnel how to collapse a parachute after an emergency landing.

Bob Davey was one of those who underwent parachute training. He recounts his experience:

> We had to learn to jump off three-foot platforms forward and backward up to five feet, higher and higher until we were jumping off sixteen-foot platforms frontward and stepping off backwards. A sixteen-foot drop was about as hard as you would hit the ground using a military parachute of that era. Much different now with the ones they've got.
>
> From there, we were attached to the cable. You climbed up the tower and hooked onto a cable, and as you came down, the instructor yanked a cord to cut you loose. You were going

down on a pulley and it dropped you to the ground. You had to hit and roll. If you put your hands up in front to stop yourself, you got two broken wrists. You had to learn to tumble forward and tumble backward. It was interesting training, but we wondered why we were jumping out of airplanes. We wanted to fly in them, not jump out of them.

CAMOUFLAGE TRAINING UNIT ADDED

A Camouflage Training Unit was added to the army air field on March 1, 1944. Ten enlisted men, working under the direction of Captain Benjamin F. Caldwell, received lectures and practical training in vehicle and aircraft concealment as well as net garnishing and folding. This was given with the idea that a unit, classified as MOS 804 Camouflage Technician, would be established. The Medical Camouflage Section worked with this detail to construct a camouflage demonstration area, but no further mention of this unit can be found.

SCANDAL ROCKS THE BASE

Seymour Sendrow served on the base in 1944. A radar supply sergeant, he recalled a scandal that rocked the base. According to Sendrow, most of the personnel from June to December 1944 had been there since the beginning of the war, enjoying the Florida weather, base amenities and military seniority. In order to maintain this lifestyle and avoid overseas duty, many of these men banded together in an underground network that secretly changed the soldiers' rotation ID numbers. Instead of rotating off the base as overseas assignments came up, their revised classification allowed them to remain behind while those who had just returned from assignments abroad were shipped out again.

"One of the radar operators had an uncle in the inspector general's office. IG came down, and within one week, we had the biggest shipping out of officers and enlisted men the base ever saw," Sendrow said.

BUILDING BOMBS

By spring 1944, scientists at Los Alamos, New Mexico, had worked out the details of the uranium 235 bomb prototype, called the "Thin Man," and began concentrating on the development of the plutonium 239 bomb. In order to launch this highly secretive mission, the military needed a variety of support personnel—pilots, mechanics, ordinance specialists and radar controllers. Command was getting geared up to select approximately 1,500 officers and enlisted men for these duties.

At the southern end of the country, in Boca Raton, troops at the army air field were getting geared up to perform duties of their own. Little did either know their paths would cross.

Paul Metro became part of the 393rd Bomb Squadron. Selected by Colonel Paul W. Tibbets, the squadron became the nucleus of the 509th Composite Group. *Photo courtesy of Paul Metro.*

COLONEL PAUL W. TIBBETS FORMS THE 509TH COMPOSITE GROUP

On June 5, 1944, one of Boca Raton Army Air Field's radar trainees, Paul Metro, was assigned to duty in the 393rd Bomb Squadron a squadron selected by Colonel Paul W. Tibbets, pilot of the *Enola Gay*. They arrived by train at Fairmont Army Air Field in Nebraska in preparation for dropping the bomb on Japan. The squadron was then ordered on September 14 to transfer to Wendover Army Air Field, Utah, where it was officially activated on December 17 and became the nucleus of the 509th Composite Group.

Metro describes this group: "The 393rd Bomb Squadron consisted of fifteen B-29 'Superfortresses,' which were modified to carry a single ten-thousand pound bomb armed with only two fifty-caliber tail guns. The crews worked with Los Alamos scientists in testing the bomb's configuration, bombing procedure and a special 'escape' maneuver after dropping the bomb. The squadron also spent six weeks in Cuba for over-water navigation."

THE BOCA RATON ARMY AIR FIELD EXPERIENCES ITS FIRST HURRICANE

While the Weather Bureau maintained a hurricane forecasting office in Jacksonville, Florida, in the early part of the twentieth century, hurricane forecasting and warning were highly ineffective. With World War II, however, military weather forecasters gained valuable experience with tropical storms, especially in the Pacific.

In 1943 Colonel Joseph P. Duckworth and navigator Ralph O'Hare flew the first intentional flight into the eye of a hurricane in the Gulf of Mexico to record wind speed, barometric pressure and direction. These reconnaissance flights with their "hurricane hunters" became crucial to the Weather Bureau in tracking storms.

Also in 1943, the hurricane forecasting office moved from Jacksonville to Miami, where the air force and navy created a joint hurricane-forecasting center. When the war ended, this center became the Weather Bureau's hurricane-forecast office and later became known as the National Hurricane Center. It was charged with the responsibility to track and forecast all tropical storms and hurricanes in the Atlantic Ocean, Caribbean Sea and Gulf of Mexico.

Meteorologists with the Pacific fleet during World War II began the practice of issuing names to tropical storms as a means of discerning one from another. Hurricanes in the Atlantic, however, were assigned numbers to identify them during any given year (e.g., the fourth hurricane of the year was referred to as hurricane number 4), but civilians remembered the storms more by the year in which they occurred, such as the 1928 hurricane.

To discern one storm from another, military weather observers begin using code names from a phonetic alphabet (e.g., Able, Baker, Charlie, etc.). This continued unofficially after the war, but was not widely publicized by the Weather Bureau until it officially began using this method in 1953.

On October 13, 1944, a storm formed in the Swan Islands in the Western Caribbean. It meandered around Grand Cayman, dropping 31.29 inches of water on the island, then tracked northward into Cuba with the eye passing just fifteen miles west of Havana. Winds were clocked at 163 miles per hour.

Forecasters in the Miami Weather Bureau followed the storm and issued warnings for the Florida Keys and southern Florida.

On October 17, 1944, base personnel were alerted that there was definite danger of a hurricane Those off base were called back to the field to make evacuation preparations. Several hundred enlisted men rushed to the Boca Raton Club to move furniture into the center of the rooms and to wash out bathtubs and fill them with fresh water.

On October 18, the entire personnel of the field moved to the Boca Raton Club with relatively little disorder or difficulties. However, upon arrival, lodging assignments for specific squadrons had not been designated, resulting in whole sections being split up.

Boca Raton Radar School described the event as follows:

> *The central area sections were scheduled to be housed on the fourth floor; however, due to more civilians arriving than had been anticipated, these enlisted men were moved to the third floor thus further crowding that floor. The 4th, 5th and 6th floors were relatively less crowded with wives and children of enlisted and officer*

personnel. A chief difficulty encountered during the hurricane was the lack of a strong central command post and lack of cooperation between various activities in the club.

No effort was made to feed hot meals, all meals consisting of C Rations. Due to the fact that most of the soldiers had never used C Rations before, there were a considerable number of minor to serious cuts resulting from their efforts to open the C ration cans. Disposal of the cans created a considerable problem. In addition, it was impossible to properly open C Rations without having some sort of container in which to put them. The men did not carry any mess gear, so a serious sanitation problem arose from the spilled C Rations plus the fact that there was no water for scrubbing or washing the floors.

Manny Chavez was assigned to assist with the evacuation. "This is interesting in that they again established the Boca Raton Club as a hurricane shelter. I say interesting again, because it had always been a hurricane shelter, the only solid building in Boca Raton in those days. Our first son, Stephen, was born at the Boca Raton 'Station' Hospital on September 6, 1944. We had to evacuate and move to the club when a hurricane hit the coast about three weeks later. By then I was a captain and assigned to the general staff of the evacuation group."

According to the *Transmitter*, two babies made their debut at the Boca Raton Club during the storm in the makeshift hospital. MPs were kept busy directing those who had lost their way in the club's labyrinth of corridors and stairways, and some in crowded rooms slept on balconies despite the intermittent drizzle.

In spite of the inconveniences, temporary overnight guests did enjoy some distraction as the 503rd AAF Band entertained them. The *Transmitter* describes the night of music: "The outfit, which brought all its instruments and two pianos (honest) with it, was broken up into one large dance band and four smaller units. Before the festivities were over, they had played on every floor."

Marion "Pint" Cornwell recalled: "There was a big band dance band. Some lieutenant was the leader and they played great music. There was a male vocalist who came out wearing a zoot suit and had quite an act."

Following notification that the storm had passed, troops moved back to the Boca Raton Army Air Field. One hundred men per floor were left at the club to clean up and dispose of all debris. Training resumed at the club on October 20.

Although Boca Raton experienced heavy rains and winds reaching sixty miles per hour, the area was mostly spared by the large storm. Coming ashore on Florida's west coast near Arcadia, the hurricane drew a vertical swath up the state to Jacksonville. At one time, its eye—shaped more like an elongated oval than a tight circle— measured almost seventy miles and stretched from Jacksonville to Ocala. Winds of fifty miles per hour or more extended two hundred miles east and one hundred miles west of the eye. Property damage around the state totaled $13 million, including 25 million boxes of fruit (very little of which was salvaged). Eighteen deaths were reported.

Lakeland citrus groves show some of the twenty-five million boxes of fruit knocked off trees by the hurricane in October 1944. *Photo courtesy of Florida's Hurricane History.*

NOTABLE EVENTS

Several events occurred on the base in 1944 that were notable enough to be mentioned in *Boca Raton Radar School*. One of them was the first Hebrew rite of circumcision: "Arrangements were made by the Jewish Chaplain for the circumcision of the infant son of an enlisted man on the Base. It was the first operation of its kind in connection with the Hebrew rite called birth milla to be performed at the Boca Raton Army Air Base."

The second was an incident that hospital personnel encountered on a routine health inspection of the base. *Boca Raton Radar School* reports the deadly May incident:

> *Easily the highlight of the month in regard to the history of the Hospital and Post in particular, and the entire State of Florida in general, was the surprise discovery of the Anopheles albimanus Wiedemann, breeding on the Boca Raton Air Field by Private Fist Class Ernest G. Erb, of the Hospital Sanitation Department. The single fourth stadium larva was obtained on 16 May 1944. On the regular weekly Mosquito Control Report this finding was confirmed by apt. Ernest R. Tinkham, Sn C., Officer in Charge of the malaria Control Program.*

Capt. Tinkham is definite in his belief that this is undoubtedly the first larva of Anopheles albimanus discovered in the Southeastern United States since George N. MacDonnell found a brood at Key West in 1904.

The full significance or importance of this scientific discovery can be readily understood when it is known that this particular species of larva is the most vicious malarial vector in the new world.

Florida borders on the southern edge of the malarial region of the country and thus there has been a relatively small amount of this particular disease found in this state. The eventual expansion of this specific larva would undoubtedly result in making Florida a highly malarious environment.

In light of this potentially dangerous situation, the Fourth Service Command Laboratory at Fort McPherson, Georgia, in conjunction with the hospital sanitation officer, conducted surveys under all bridges and in tunnels. No adult larvae were found; however, the United States Public Health Service initiated a thorough spray program.

Another concern by medical personnel in May was the increase of venereal disease among the troops as noted in *Boca Raton Radar School*: "The overall rate of 30 per 1,000 shows an increase of 17 per 1,000 over the previous April." No valid explanation was offered as mandatory lectures were given twice weekly by hospital personnel. By August, the rate dropped.

NOTABLE ENTERTAINERS

Many entertainers joined the army during the war to support the effort and demonstrate their patriotism. While most weren't on the front lines, they played an important part in raising morale of the troops wherever they were stationed.

Staff Sergeant Tony Martin, a popular singer and recording artist, was one of the more prominent servicemen to be stationed at the Boca Raton Army Air Field. He served on the base as co-master of ceremonies of the weekly soldier show, a variety show comprised from the ranks of the enlisted men, and Martin entertained troops by crooning such favorites as "Sweet and Lovely" and "Begin the Beguine."

Private Benjamin "Benny" Payne was a notable black pianist stationed at the Boca Raton Army Air Field. Staying around his hometown of Philadelphia, Pennsylvania, until he was twenty-one, he went to New York and landed a job with the hit show *Blackbirds of 1929* that went to Europe the following year. Next, he played with Connie's Chocolates that starred Cab Calloway as leading man and singer. Payne followed that show with a job at the Cotton Club.

When he came to Boca Raton after just nine months in the army, Payne had previously been playing with the Cab Calloway band for twelve years. According to the *Transmitter*, Payne "stopped the show" with his boogie-woogie performance at the special services show.

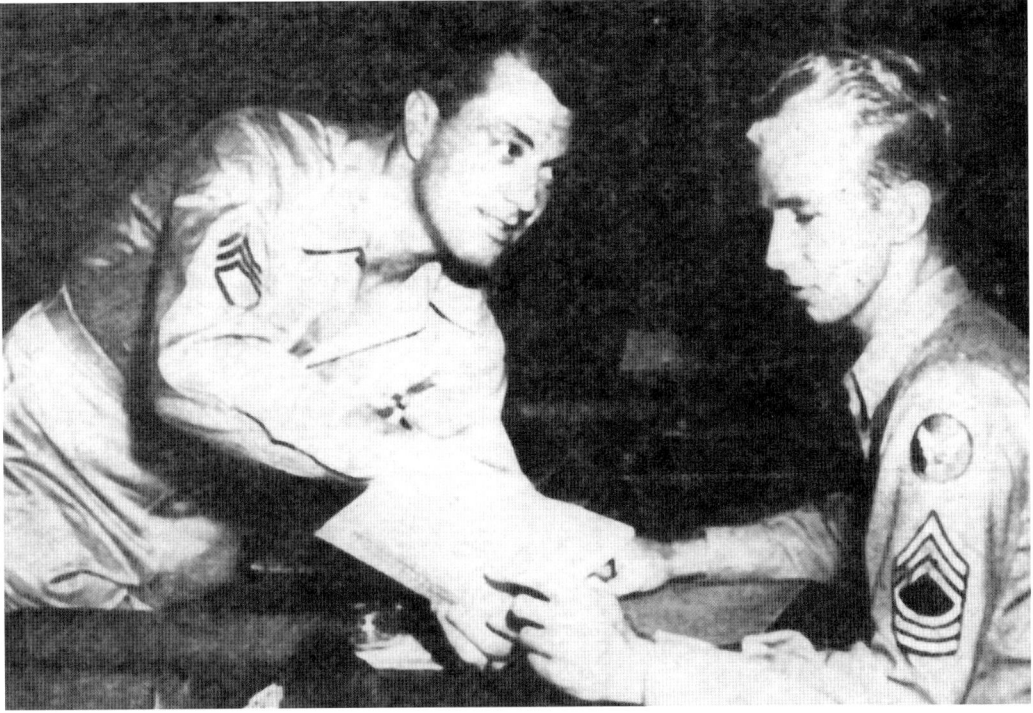

Popular singer Tony Martin served in Special Services. He hosted weekly variety shows and entertained troops. *Photo courtesy of the Boca Raton Historical Society.*

Another talent to join the ranks at the Boca Raton Army Air Field was twenty-one-year-old Corporal Theodore E. Gurtner, the voice of Walt Disney's famous character Donald Duck. Before army duty, Gurtner was a narrator for the Walt Disney Productions for five years.

He entertained troops on and off the base with his imitations and read fan mail that seemed to follow him wherever he went. When he discovered some of his barrack-mates didn't receive mail, he started handing out his own mail, letting the soldiers read the letters first.

Staff Sergeant George Lex—part of the artist's staff of Fleischer Studios, which made feature-length and short cartoon movies such as *Gulliver's Travels*, *Popeye* and *Superman*—drew posters with messages against the Nazis and Japanese to help the military educate GI Joes on the danger of the enemy. He also made signs for the base and was a member of the Boca Raton field band.

Charlie Banks, a dancer, had six hundred flying hours to his credit before becoming a flight instructor with the War Training Service at Tuskegee in 1942. When the program was abandoned, Banks became a private. After completing a radio course at Scott Field, he was sent to the Boca Raton Army Air Field to help the base's entertainment program.

Left: Private "Benny" Payne, who played with the Cab Calloway Band, stopped the show with his boogie-woogie performance. *Photo courtesy of the Boca Raton Historical Society.*

Right: Sergeant Jack Lacey played with the 503rd AAF Band. *Photo courtesy of the Boca Raton Historical Society.*

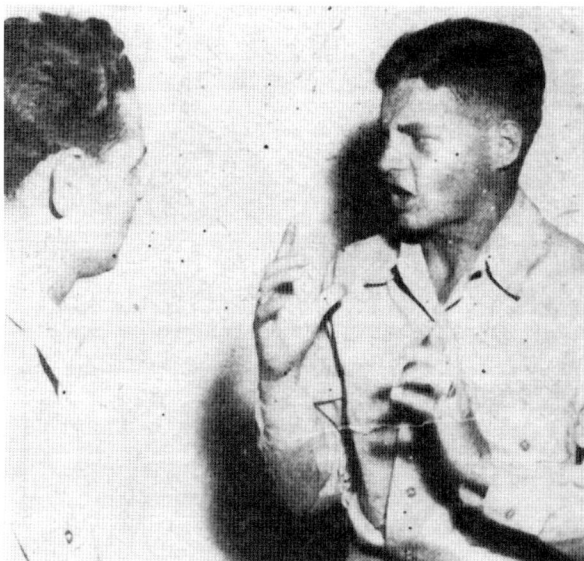

Left: Theodore Gurtner was the voice of Disney's famous cartoon character Donald Duck. *Photo courtesy of the Boca Raton Historical Society.*

Right: A quick-stepping hoofer, Charlie Banks danced with the likes of Fred Astaire, Josephine Baker and Cab Calloway. *Photo courtesy of the Boca Raton Historical Society.*

Artist George Lex drew posters with messages about the Nazis and Japanese to help educate troops about the enemy. *Photo courtesy of the Boca Raton Historical Society.*

Sergeant Jack Lacey played twenty years with top-notch bands. Andre Kostelanetz called him the "greatest" trombonist in the nation.

BOCA RATON CLUB RETURNED TO CIVILIAN CONTROL

J. Meyer Schine purchased the Boca Raton Club from the G.H. Geist Trust, and in December 1944, the club returned to civilian control. Tourists once again enjoyed the exquisite ambiance and impeccable service of the elegant club, something they had sorely missed.

CLOSING OUT 1944

The year 1944 brought growth and changes to the Boca Raton Army Air Field. Thousands of troops were educated and sent abroad, and fresh recruits arrived to occupy their vacated bunks. New programs were added to the training curriculum, and diversions on and off the base occupied the leisure time of the men and women who faithfully served their country.

At MIT, the Rad Lab had given birth to the Manhattan Project. It was now poised to deliver a deadly blow to the isle of Japan.

1945 Time Capsule

January—Pepe LePew debuts in the Warner Bros. cartoon *Odor-able Kitty* • A U.S. Army bomber crashes into the seventy-ninth floor of the Empire State Building • Jackie Robinson signs with the Montreal Royals • The microwave oven is patented

World News

• Italian partisans kill Prime Minister Benito Mussolini.
• World War II comes to an end—May 8 in Europe and August 14 in the Pacific—after involving fifty-seven nations and leaving 55 million people dead.
• Soviet troops liberate prisoners of the concentration camp at Auschwitz. An estimated 6 million people died in the German camps.
• The United Nations convenes with representatives from 51 nations.

National News

• Harry Truman takes the office of president after Franklin D. Roosevelt passes away.
• Japanese-Americans are permitted to return to the West Coast.
• The United Auto Workers strike against General Motors for 113 days.
• Grand Rapids, Michigan, is the first U.S. community to fluoridate water.
• Five thousand homes have TV sets.
• Ball-point pens, Tupperware and frozen orange juice all hit the market.
• U.S. Army General George S. Patton passes away.

Sports

• World Series champion—Detroit Tigers • U.S. Open golf champion—Was not held in 1945 • Pro football champion—Cleveland Browns • Indianapolis 500 winner—Was not held in 1945 • Stanley Cup winner—Toronto Maple Leafs • NCAA basketball champion—Oklahoma A&M • College football champion—Army • Heisman Trophy winner—Felix "Doc" Blanchard from Army

Music

• "Rum and Coca-Cola"—The Andrews Sisters
• "Sentimental Journey"—Les Brown
• "There, I've Said it Again"—Vaughn Monroe
• "Till the End of Time"—Perry Como
• "White Christmas"—Bing Crosby
• "It's Been a Long, Long Time"—Harry James
• "Accentuate the Positive"—Johnny Mercer
• "Candy"—Johnny Mercer and Jo Stafford
• "Chickery Chick"—Sammy Kaye
• "Dream"—Pied Pipers

Movies

• *The Lost Weekend*—Academy Award winner
• *Spellbound*
• *And Then There Were None*
• *A Tree Grows in Brooklyn*
• *The Bells of St. Mary's*
• *Detour*

Cost of Living

New house	$4,625.00
Average income	$2,390.00 per year
New car	$1,025.00
Average rent	$60.00 per month
Tuition at Harvard University	$420.00 per year
Movie ticket	$.50 each
Gasoline	$.15 per gallon
U.S. postage stamp	$.03 each

Food	
Granulated sugar	$.75 for 10 pounds
Vitamin D milk	$.62 per gallon
Ground coffee	$.50 per pound
Eggs	$.22 per dozen
Fresh baked bread	$.09 per loaf

1945

BOCA RATON ARMY AIR FIELD MIRED IN CHALLENGES

While Allies fought battles in Europe, the Boca Raton Army Air Field had battles of its own. By 1945, 16,742 men were stationed at the base and tens of thousands more had passed through the front gate, completed radar training and shipped out to Europe or the Pacific. While this brought about positive results in the war effort, it proved a hindrance to training as the quality of trainees declined. According to *Boca Raton Radar School*: "If student quality was based on aptitude to absorb radar training, mentally, or on GOT scores the quality had steadily declined since the beginning of the school. During the early stages of the school, and the war, the manpower barrel was full. Men with serious interest in electronics, or with civilian technical radio experience, were carefully selected to study radar. As the War matured, the manpower supply decreased, and the quality of students declined."

Low student quality wasn't the only problem facing the army air field. The rapid transitioning of troops on and off the base negatively affected entire operations. So quickly were these changes made that it became impossible for the personnel section to keep up. The processing of records sorely lagged behind arrivals, resulting in chaos in the assignment of housing and training schedules. Additionally, payroll glitches caused some troops to go unpaid for months. On one such occasion, the Red Cross stepped forward and loaned GIs $10 so they could purchase Christmas presents for their families.

Not only were new recruits arriving at the Boca Raton Army Air Field, but as the war wound down, troops returning from Europe arrived as well. Many of the jobs that were filled by these troops during the war no longer existed. It became necessary to retrain these men, but this was slow to materialize, and troops found themselves in limbo. Pilots who had fulfilled their wartime flying duties were no longer needed and found

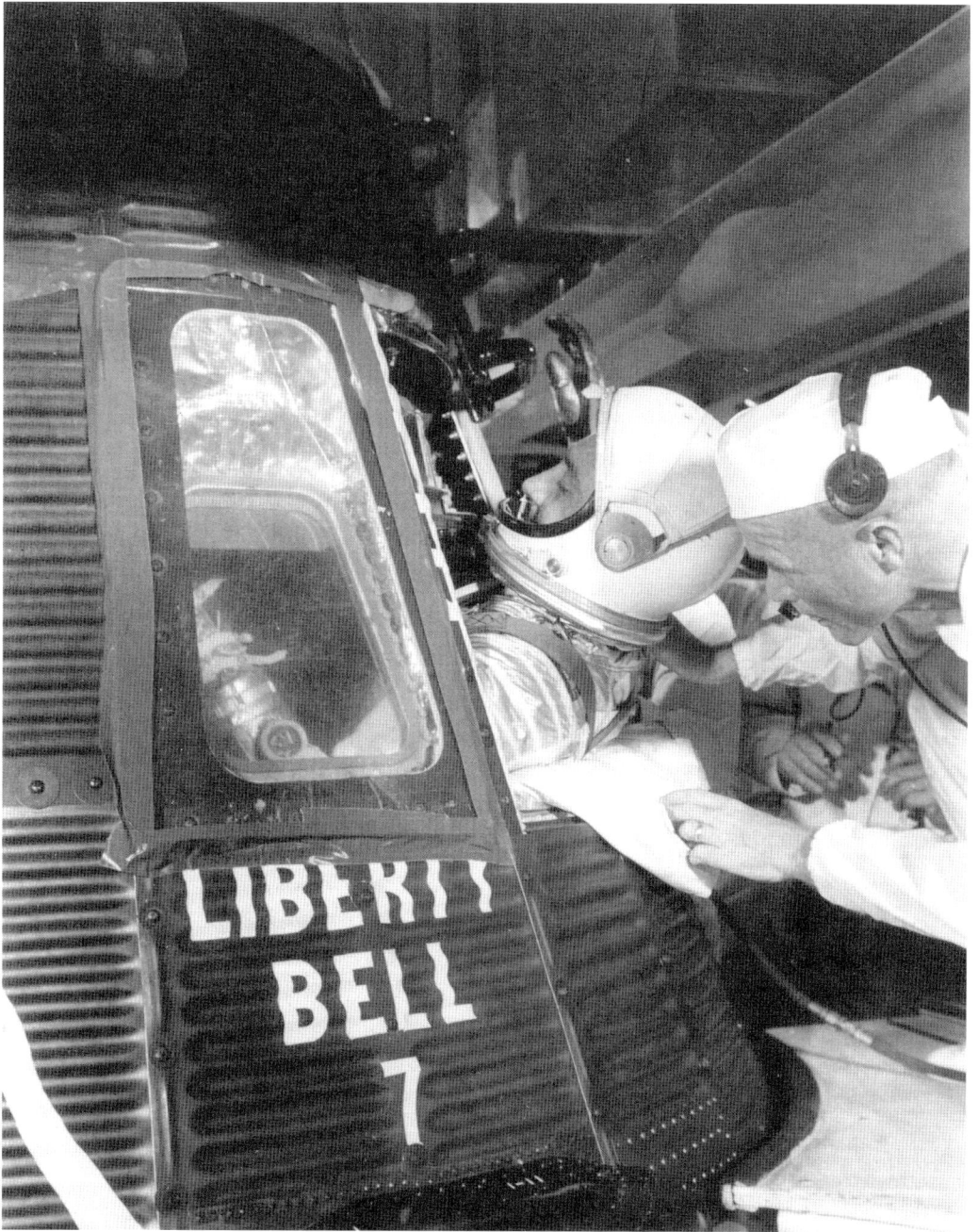

Before becoming an astronaut, Gus Grissom (left) served at the Boca Raton Army Air Field as a clerk typist. *Photo courtesy of NASA.*

their flight pay terminated. This resulted in a reduction of income that exacerbated an already frustrating situation.

Instructors were reassigned at a moment's notice. Along with them went their wives, who served vital support roles throughout the base. There were never enough sales clerks in the PX. Clerk typists became a scarce commodity. Troops were reassigned or transferred to fill these jobs, but there never seemed to be enough personnel to meet demand.

One clerk typist who served on the base later became one of the nation's pioneer astronauts.

LIEUTENANT COLONEL GUS GRISSOM AT BOCA RATON

While he soared to great heights later in his career—piloting the *Liberty Bell 7* spacecraft in 1961 and the first manned *Gemini* flight, a three-orbit mission in March 1965—Lieutenant Colonel Virgil "Gus" Grissom's time at the Boca Raton Army Air Field wasn't at all glamorous.

Following high school graduation in 1944, Grissom, a native of Mitchell, Indiana, joined the army corps and went to Wichita Falls, Texas, for five weeks of basic training. In January 1945, he reported for duty at Boca Raton as a clerk typist. Briefly taking time off in July to return home to marry Betty Moore on July 6, he then returned to Florida. His hope was to receive pilot training; however, that never occurred and he was discharged in September 1945.

He later received a bachelor of science degree in mechanical engineering from Purdue University and decided to rejoin the armed services. He became an air cadet at Randolph Air Force Base in Texas and received his wings in March 1951. Grissom flew one hundred combat missions in Korea in F-86s with the 334th Fighter Interceptor Squadron and became a jet instructor at Bryan, Texas, in 1952.

In August 1955, he entered the Air Force Institute of Technology at Wright-Patterson Air Force Base, Ohio. There he studied aeronautical engineering. In 1956, he attended test pilot school and the next year returned to Wright-Patterson as a test pilot.

Russia shocked the world in 1957 by announcing the successful launch of its first satellite, Sputnik. After that, the race into space moved rapidly. Seven jet pilots were selected in 1958 to serve as astronauts in America's first man-in-space program, Project Mercury. Among those selected in 1959 were John Glenn, Walter Schirra, Alan Shepard, Scott Carpenter, Donald "Deke" Slayton, Gordon Cooper and Gus Grissom.

After two successful missions into space, Lieutenant Colonel Grissom died on January 27, 1967, at the age of forty aboard the *Apollo* spacecraft. A flash fire during a launch pad test at the Kennedy Space Center in Florida killed Grissom, Edward White and Roger Chaffee.

On July 4, 1945, the long-awaited officer's club opened amid much fanfare. *Photo courtesy of the Boca Raton Historical Society.*

Officers' wives enjoy the annual luncheon hosted by the wife of the post commander. *Photo courtesy of the Boca Raton Historical Society.*

LONG-AWAITED OFFICER'S CLUB OPENS

From 1942 to 1944, the Boca Raton Club served admirably as the temporary home of the officer's club. But when the Boca Raton Club returned to civilian authority in December 1944 and officers took up residency on the base, a new officer's club became a necessity.

On July 4, 1945, the new facility opened amid much fanfare. A reception hosted by top brass was held for officers and their wives. Featured was dinner and dancing with music by members of the 503rd AFF Band.

CUBAN NIGHTS

Officers enjoyed spending after hours at the officer's club, but when they made duty flights to the Caribbean, nightlife took on a whole new meaning.

Twenty-year-old Charles R. Bender arrived at the Boca Raton Army Air Field in the spring of 1945. He recalled trips to the Caribbean:

Charles Bender in the cockpit of an AT-6 "Texan," an advanced trainer. *Photo courtesy of Charles Bender.*

I remained at BRAAF doing routine radar work and flew as often as I could. On three occasions, we flew a B-17 to Batista Air Field, outside Havana, [Cuba,] where we enjoyed a couple [of] days of sightseeing and recreation. The nightlife was incredible, much like Las Vegas is today. Nothing shut down. During this time, after an all-night party, I watched armored vehicles pull up to the Hotel Nacional and Cuban soldiers carry out many large duffel bags and load them into their vehicles. Later we found out that the bags contained money from the gambling casino and were being taken to General [Fulgencio] Batista.

Before leaving Cuba, we loaded cases of scotch whiskey, cartons of cigarettes and cases of Havana cigars onto our plane and brought them back to BRAAF. On one occasion, we flew this type of cargo to West Point, New York, landing there at Stewart Army Air Base.

OFF-BASE ENTERTAINMENT

While planes were flying to Cuba and base operations were mired in confusion, troops managed to attend training classes and perform necessary on-post duties. When off duty, they were given passes so they could enjoy South Florida's nightlife and entertainment.

Two commodities dictated the extent of a soldier's leisure time: spending money and transportation. Bob Davey recalled: "There wasn't an awful lot we could do because you can't go very far on less than $50 a month minus the cost of your insurance. Later on, when we finished the last six-week course in radar, we got corporal stripes, and we felt filthy rich because we went from $50 a month to $66 a month."

On weekends, GIs gathered at their squadron and loaded into long open trailers with benches running down both sides and in the middle. Towed to the bus station on U.S. Route 1, they jammed both sides of the highway, waiting for transportation. Those too impatient to wait for the bus flagged down passing motorists in hopes of hitching a ride. Their intended destination was either north to Delray Beach and West Palm Beach or south to Fort Lauderdale and Miami.

Philip Kantor was one of the soldiers who took weekend jaunts to Miami Beach. "I considered Miami Beach 'heaven,' as a nineteen-year-old away from home for the first time. My friends and I went to the Carlyle Hotel on Ocean Drive each time we were at the beach, having established a firm relationship with the desk clerks."

Nelson Godett found residents of Florida's hottest nightspots "very receptive" to servicemen. John Cochrane noted: "For one dollar a serviceman or -woman could get an air-conditioned room in some of the Collins Avenue hotels. And even though the whole of South Florida area teemed with service personnel, the locals were exceptionally nice to us. I don't recall one negative incident."

Instead of going south like many of his friends, Ed Malavarca usually went north. "We went up to West Palm Beach where there happened to be two camps for women," he said, "one for the WACS (Army) and the other for the SPARS (Navy). So we went up there to meet the girls at the back gate. We socialized with them, nothing serious, then we went home."

Louis Delgado went farther north, to Morrison Field in West Palm Beach. He recalled an unusual experience:

> I had a friend who was stationed at the Air Transport Command's Morrison Field in West Palm Beach. Whenever he was away on a pass, I would sleep in his bunk so I could spend time either at the beach or walking around Palm Beach, admiring the large mansions or the boats anchored on the Intracoastal Waterway. I remember one time when I was sleeping in my buddy's cot, I was awakened at about three or four in the morning and I was told to get up for KP duty. I said I wasn't from that base and I would not get up. Fortunately for me nothing happened, and I just went back to sleep. A towel was tied to the foot of the cot to indicate who was scheduled for KP, and my buddy either forgot that he was scheduled or he just went off base regardless, so they thought that I was him.

Late at night when the fun was over, troops repeated their earlier performance by congregating at bus stops to return to the base. Those unable to catch the last bus at 2:00 a.m. were on their own and either had to hitch a ride or hoof their way back to the base. Bill Robison remembered such an experience:

> The stretch from Delray to Fort Lauderdale at night was one of virtual total darkness. I well remember missing the last bus out of Miami (the midnight bus) and having to hitch hike back to Boca. Since there was very stringent gas rationing, this was not easy to do. However many of the permanent party members at BRAAF were allotted a certain ration of gasoline, and they were more than willing to give a G.I. a ride. However, car after car would come by, stop momentarily and the driver would yell out "sorry buddy, but I have a full load." Full indeed, there were bodies upon bodies of GIs stacked like cordwood in those old sedans.
>
> The other hope for a ride was with commercial fishermen mostly out of the Deerfield or Pompano area who were allotted gasoline to support this critical aspect of the war effort. However if you got a ride with them you were let out on the highway somewhere in this deserted stretch. What little habitation might have existed was near the water a considerable distance from the highway.
>
> I remember standing literally for hours in total darkness and listening with considerable apprehension to the cries and calls of animals in the Everglades, which at that time extended virtually to the highway. Occasionally a pinpoint of light would appear down the highway. Most of the time it was a car stacked with GIs with room for not one more. Presently, it would be daybreak and not long after there would be cars coming up the highway heading for BRAAF. These would be, for the most part, civilian employees who commuted from the Ft. Lauderdale area. Thus, gratefully, I never missed Monday morning roll call.

The provost marshal's office at the main gate. *Photo courtesy of the Boca Raton Historical Society.*

TROUBLE IN PARADISE

With Fort Lauderdale, Miami and Palm Beach jammed with weekend military personnel, problems were bound to arise. Initially, military police dispatched to Fort Lauderdale rode with civilian police to help with problems and the arrests of troops. Later, however, when the Fort Lauderdale Police Department began to try to send military personnel to civilian courts, the Boca Raton Army Air Field provost marshal, who believed these men should go through the military justice system, ordered MPs to cease riding in civilian police cars. Two days later, the old system was back in place.

Military police were ordered to pull out of the city by 0200 hours (2:00 a.m.), but numerous reports of trouble occurred after that time, especially at the bus stops. With long lines of tired and tipsy soldiers waiting for rides back to the base and overcrowded transportation, the presence of MPs was necessary. In July, MPs pulled all night duty.

MPs were called upon to assist civilian police in monitoring off-duty soldiers in Fort Lauderdale.
Photo courtesy of the Boca Raton Historical Society.

While white troops may have paid for being drunk and disorderly by being arrested, black enlisted men incurred much worse punishments. *Boca Raton Radar School* noted: "[Black] EMs have been beaten by civilian police in the presence of Military Police for not saying 'Sir' or removing their hats while talking with 'whites' or for failure to admit on the order of civilian police that they are a 'Nigger.'"

To their credit, commanding officers at the Boca Raton Army Air Field recognized problems arising from segregation and took steps to correct them both on base and with civilian authorities. The solutions included giving African-American enlisted men an additional hour before bed-check, discussions regarding rights of African-American soldiers with civilian police departments, eliminating segregation in the base theater and giving black EMs better seating arrangements. While these may not seem big steps by today's standards, they were monumental in those days.

Discrimination didn't begin and end with the black troop population, gays were also targeted for discrimination by some military personnel. A young sixteen-year-old Ned Renner remembered his eye-opening encounter with prejudice exhibited by base personnel: "Two of the students in my radar class always seemed to have excessive amounts of spending money. More than army privates usually have. I questioned one of them one day, and he not only told me where they were getting their money but invited me to go with them on weekends. I learned that the two of them were going to Miami where they would beat and rob homosexuals. I immediately refused their invitation and severed all contact with them, knowing this was illegal and immoral, even at my young age."

BEACH ENJOYED BY TROOPS

For soldiers who preferred to stay close to the base, many enjoyed the Boca Raton beach, which was accessible from Palmetto Park Road. Nelson Godett remembered his trips to the beach: "If we didn't take off to Miami Beach, we could go over to the beach across the Intracoastal [Waterway] and actually go out and swim in the ocean. And the interesting part was a lot of the fellows were from the mountains or the Midwest, and had never seen salt water. They'd go running down the beach and dive in the water and come up sputtering something while the rest of us laughed our heads off at them because, I myself, being born about one hundred yards from Long Island Sound, was used to the salt water and they certainly weren't."

Ned Renner enjoyed the beach as well, although he remembered a lot of sand crabs. Sometimes he would take a bayonet with him and, on the way to the beach, pick up coconuts to snack on. "They were plentiful and lay on the ground under the coconut trees," he said. Unfamiliar with the Florida sun and eager to get a tan, he attempted to get it instantly. When he left the beach, he was beet red.

"The following day, blisters began to appear all over my body, many of which were larger than golf balls. I went on sick call because of the pain and blisters. I

Charles Bender poses by a palm tree at Boca Raton beach. *Photo courtesy of Charles Bender.*

Soldiers enjoy the beach pavilion at Palmetto Park Road and A1A. *Photo courtesy of the Boca Raton Historical Society.*

was given what I believe may have been calamine lotion and a written medical excuse giving me permission to attend classes without a shirt. I stood out among the troops. It was a very painful experience, and I could have very well been court-martialed," said Renner.

When transportation wasn't available, troops hoofed it to the beach. "The walk to the beach was a pleasant one along palm-lined Palmetto Park Road," said Ed Wallish. "There were some impressive houses along the street. One house I particularly remember was a pink stucco Georgian style house surrounded by a pink stucco wall; the wall had a small metal plaque that proclaimed the house to be the 'Casa Rosa.' Where Palmetto Park Road crossed the waterway there was a wooden drawbridge where I watched the translucent jellyfish pulsing along on their journey."

In the summer of 1945, school sessions, unless taught in the hangars, were conducted very early or at night so as to avoid the midday heat. According to John H. Cochrane, that meant plenty of beach time: "I still remember most distinctly walking one day to the beach through a large uncompleted subdivision on the east side. All the concrete streets were in, as were all the streetlights, but the jungle had moved back in and now covered most everything. Because I was a car buff, even back then (like most boys), I took particular note of an early 1920s Cadillac sedan back in the weeds. It was up on blocks alongside some old, decrepit house. The whole scene seemed, to this nineteen-year-old, a living testimony to the greed and the financial folly of that earlier time."

UNITED SERVICE ORGANIZATION (USO)

Troops enjoyed United Service Organization (USO) shows or other forms of entertainment while on the base. Paul Metro went to USO shows quite regularly. "At Boca Raton, we saw several USO shows, one with Leo Durocher, manager of the Brooklyn Dodgers, and Danny Kaye, who was an up-and-coming comedian at the time. Other shows I remember included a variety show featuring 'Miss Gams,' a pretty girl with long legs. Out at the Boca Raton Club, champion golfer Gene Sarazen held a trick-shot exhibition."

Other activities included watching movies in the base theater or socializing with local girls. Ned Renner recalls how he spent his leisure time:

> *I became friendly with some young ladies from Hialeah [Florida], one which was the daughter of parents who owned some riding stables. One Saturday night, instead of meeting at the USO club, a group of us met at her house, where we mounted up on horses and rode to an unfamiliar bar in the area. I recall there being a place in front of the bar where we tied up the horses just like the old Wild West. We left the bar several hours later and decided to race back to the stables. Unfortunately, my horse stumbled in the sand, and I was thrown off into the sandburs below. We were unable to catch the horse, and I rode double back to the stables. I do not remember any of these ladies by name. However if any of them are living today, they would remember that event.*

Cartoon artist Herb Roth gives his impression of troops enjoying a night out. *Photo courtesy of the Delray Beach Historical Society.*

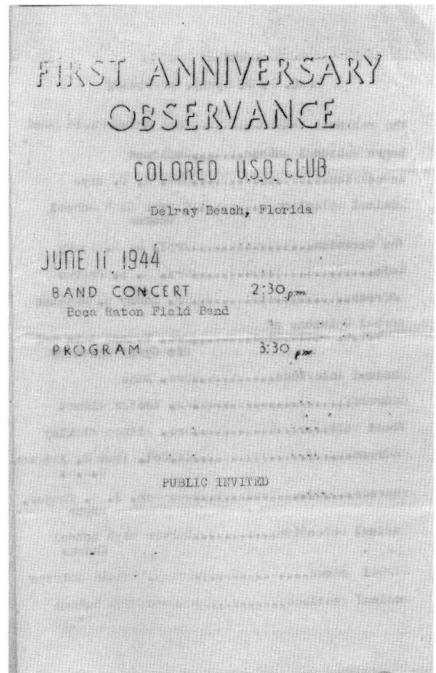

On July 11, 1943, a colored USO club opened in Delray Beach. A year later, an anniversary celebration was held with entertainment provided by the 503rd AAF Band. *Photo courtesy of the Boca Raton Historical Society.*

GI University Established

To keep GIs interested in higher learning, a partnership between the Boca Raton Army Air Field and the University of Miami was pursued to offer college courses for credit. After the university felt the educational level of the men did not warrant university participation, the base set up its own school, GI University.

Courses included art, math, Spanish, physics, business management, news reporting and salesmanship. A total of 397 GIs enrolled in the first courses. By the second semester, more than 700 GIs were enrolled in twenty courses, but by the end of the year, so many troops shipped out that only 111 students remained.

Victory in Europe: V-E Day

In April 1945, while troops at the Boca Raton Army Air Field enjoyed off-duty time and attended classes, the Allies in Europe were overrunning German troops from the west while Russian forces advanced from the east. On April 25, the last of the strategic targets, the Skoda armament works at Pilsen, Czechoslovakia, was bombed, allowing the Army Air Force to turn its attention to missions of mercy. Pilots dropped food relief in northern Italy and the Netherlands, and evacuated prisoners of war.

Five days later, Adolf Hitler committed suicide in his Berlin bunker. Taking command of the military, Admiral Karl Doenitz, sent General Alfred Jodl to the Supreme Headquarters Allied Expeditionary Forces detachment in Rheims to seek terms for the war's end.

On May 7 at 2:41 a.m., General Jodl signed for unconditional surrender of German forces on all fronts, which was to take effect on May 8 at 11:01 p.m. After six years and millions of lives lost, the Nazis were crushed and the war in Europe finally over.

Throughout America, the joyous news of the end of the war in Europe brought celebration. At the Boca Raton Army Air Field, the momentous occasion was marked by a somber service in the chapel. But the war was only half over. Troops in the Pacific still fought the Land of the Rising Sun, and it would be three more months before World War II finally came to an end.

Enola Gay Makes History

While troops at the Boca Raton Army Air Field engaged in everyday activities, the members of the 509th Composite Group prepared for the most important mission of their lives—one that changed the course of the war and history.

Paul Metro, a Boca Raton student and radar trainee, told of his experience as a member of this group:

In April 1945, the group was ready to move overseas. Some traveled by air, but most by sea. The ground echelon left Seattle May 6 and arrived on Tinian in the Marianas on the 30ᵗʰ of May, settling in quarters near North Field.

Our airplanes had familiarization missions and on 20 July flew its first mission over Japan. These missions were not part of the 100 plane missions flown from Guam, Saipan and Tinian and the group took a razing for this. Our planes had tail markings of other units and carried a large bomb, painted orange, called a "Punkins." The object was to lull the Japanese into thinking they were reconnaissance missions or that they got separated from their units. Radar was used for bombing in 30 percent of these missions.

On August 5, 1945 the group was ordered to carry out the mission it had trained for. The Enola Gay *left North Field at 2:45 AM the next day and dropped "Little Boy," a uranium bomb on Hiroshima. That morning, on Tinian, was when we first learned that it was a nuclear bomb. We celebrated the "end of the war." It proved to be premature.*

Japan would not surrender, so a second atomic mission was ordered three days later. The primary target was Kokura, but due to overcast skies, visual bombing was the order and the plane flew on to Nagasaki where "Fat Man," a plutonium bomb, was dropped. This was the last nuclear mission, however, conventional bombings continued until August 14 when Japan accepted the Potsdam Ultimatum.

Victory in Japan: V-J Day

Nelson Godett was enrolled in the Air Observer School learning how to bomb from B-24s when word came of victory: "I remember being at a weather briefing at 1900 hours on August 14ᵗʰ prior to a night training mission when we were informed the war was over. Our mission was cancelled, and we went to the officer's club to celebrate."

Marking the end of the war, chapel services were held to give thanks for peace and Sunday services were filled to capacity when the day was set aside as a "National Day of Prayer." Troops enjoyed a two-day holiday, and pilots were relieved of flight status for six days.

Commanding officers expected that after the break, it would be back to work as usual. But that was not the case. Minds that once concentrated on radar mechanics now concentrated on: "When am I getting out?" With the war over, civilian women felt they no longer needed to volunteer—resulting in an even more severe shortage of support personnel than normal.

The Army Air Corps Technical School of Radar was in a "very bad state," as physically qualified personnel were shipped out. In a twenty-eight-day period, the school lost 700 instructors, maintenance men and supervisors, and 140 others were sent to fill clerk, cook and MP positions where there was a critical shortage. *Boca Raton Radar School* stated: "Very, very few of the students are the least bit interested in radio or radar." The solution was to reduce the student population by 50 percent and transfer them to other positions.

SEPTEMBER 15 HURRICANE

The hurricane started in Cape Verde and marched westward through the Bahamas. Eventually, it came ashore in Florida on September 15 just below Homestead and cut a forty-mile-wide swath south of Miami. Wind speeds were clocked at 138 miles per hour at Carysfort Reef Lighthouse and 107 miles per hour in Miami. However, records at the U.S. Naval Air Station at Richmond Heights in Miami indicated an average two-minute maximum wind velocity of 170 miles per hour, and it was estimated that the wind reached 196 miles per hour for a few seconds.

At the naval station, hangars collapsed during the climax of the storm and enormous fires erupted. Fueled by high-octane gasoline and fanned by ferocious winds, the base lost 25 blimps, 183 military planes, 153 civilian planes and 150 automobiles. More than two hundred people were injured in the blaze.

The hurricane hit the Boca Raton area shortly thereafter. Nelson Godett was on base at the time: "B-24s were being flown to the Midwest for safety. Never having experienced a hurricane, I opted to remain on base and was assigned to install shutters on the existing barracks. With the help of one hundred enlisted men, the carpentry shop and one two-ton truck, we made and installed 1,110 shutters."

After making landfall, the storm turned almost due north, drew a line up the Florida Peninsula and curved back out to sea at the Georgia line.

RAD LAB CLOSES

At the conclusion of the war, an open invitation went out to members of the press, inviting them to tour the Rad Lab. What was once a top-secret operation was now to become an open book. But before all the invitations were delivered, news of Japan's surrender hit the streets.

Jennet Conant in her book *Tuxedo Park* tells what happened next:

Time magazine's scheduled cover story on radar was bumped to page 78, and the new cover, celebrating V-J Day, credited the work of the Los Alamos physicists. The men who worked on the atomic bomb were hailed as heroes, and countless books and Hollywood movies would recount their exploits, while the daring and inventive minds who created radar were largely forgotten.

The Manhattan Project became world famous. The Tizard Mission faded into obscurity. Only the Rad Lab veterans knew better—knew that if radar had not kept the Germans from defeating England, the war might have been over before America entered the contest. Everyone who had worked at the laboratory understood the decisive role their deadly devices had played in speeding the day of victory, and it was reflected in a remark by Lee DuBridge [who ran day-to-day operations at the Rad Lab] *that became something of an unofficial slogan, their badge of honor:* "*Radar won the war; the atom bomb ended it.*"

The Rad Lab at the Massachusetts Institute of Technology delivered the decisive blow to the German U-boat and shot down German planes and V-1s [the first modern guided missile used in wartime, commonly called a buzz bomb]. It introduced many revolutionary concepts into warfare and gave birth to a multibillion-dollar industry, soon to be absorbed by private enterprise.

On December 31, 1945, the world's most important wartime research and development facility slipped silently into history amid the same quiet fanfare with which it had emerged. Most of the physicists returned to their former university jobs, resuming their careers as professors and research scientists.

CLOSING OUT 1945

The war was over, and a transition period was about to envelope the Boca Raton Army Air Field. With flight training discontinued, many soldiers were discharged. Combat and noncombat aircraft, declared excess property, were disposed of at salvage depots. On base, there appeared to be an excess of everything from bedding to gas masks. Now that husbands were home and job security on the base was questionable, civilian employees resigned, resulting in low morale.

In December, Brigadier General Winslow C. Morse became post commander as the military complex nationwide went through post-war reorganization. The Boca Raton Army Air Field was about to undergo major changes.

1946 TIME CAPSULE

The first radar contact with the moon is made by the U.S. Army Signal Corps • The first electric blanket is sold for $39.50 • The first American-built rocket leaving Earth's atmosphere reaches an altitude of fifty miles • The Flamingo Hotel opens in Las Vegas, Nevada, marking the beginning of an era

WORLD NEWS
• Europe and Japan begin the rebuilding process.
• English Prime Minister Winston Churchill gives his "iron curtain" speech.
• Lebanon declares its independence from France.
• The United States conducts atomic tests on Bikini Atoll, in the Marshall Islands.
• The United Nations holds its first session.
• The bikini swimsuit, named after Bikini Atoll and the atomic explosions released there, debuts in Paris, France.
• The People's Republic of Albania is established.

NATIONAL NEWS
• The U.S. Atomic Energy Commission, a civilian group, is formed by Congress.
• The Central Intelligence Agency is formed.
• World War II veterans, making use of the provisions of the GI Bill of Rights, head to college in record numbers.
• Birthrates jump to more than 1.4 million in one year with the return of World War II veterans.
• The United States gives the Philippines independence.
• AT&T announces the first car phones.
• Alaskans vote in favor of statehood.

SPORTS
• World Series champion—St. Louis Cardinals • U.S. Open golf champion—Lloyd Mangrum • Pro football champion—Chicago Bears • Indianapolis 500 winner—George Robson • Stanley Cup winner—Montreal Canadiens • NCAA basketball champion—Oklahoma A&M • College football champion—Notre Dame • Heisman Trophy winner—Glen Davis from Army

MUSIC
• "The Old Lamplighter"—Sammy Kaye
• "Ole Buttermilk Sky"—Kay Kyser
• "Personality"—Johnny Mercer
• "Rumors Are Flying"—Frankie Carle
• "Surrender"—Perry Como
• "Symphony"—Freddy Martin
• "To Each His Own"—Ink Spots
• "I'm a Big Girl Now"—Sammy Kaye

MOVIES
• *The Best Years of Our Lives*—Academy Award winner
• *Notorious*
• *My Darling Clementine*
• *Song of the South*
• *The Big Sleep*
• *It's a Wonderful Life*

COST OF LIVING

New house	$5,600.00	
Average income	$2,500.00 per year	
New car	$1,125.00	
Average rent	$65.00 per month	
Tuition at Harvard University	$ 420.00 per year	
Movie ticket	$.55 each	
Gasoline	$.15 per gallon	
U.S. postage stamp	$.03 each	

Food

Granulated sugar	$.75 for 10 pounds
Vitamin D milk	$.70 per gallon
Ground coffee	$.50 per pound
Eggs	$.22 per dozen
Fresh baked bread	$.10 per loaf

1946

CHANGES ARRIVE

Brigadier General Winslow C. Morse recognized the lackadaisical attitude that permeated the Boca Raton Army Air Field and decided a stricter hand was necessary to refocus the soldiers on post-war objectives even though no one was certain just what those were.

Military command deactivated the Women's Army Auxiliary Corps as well as drew up plans to consolidate operations and personnel into one section of the base In addition, because of the laissez-faire attitude that had permeated the base, security tightened as the base adopted a more restricted and unified pass system, requiring the listing of names of the men leaving the base during duty hours. It also called for the checking of all on-base passes and patrolling cities within a twenty-five-mile radius for men without passes, leave or furloughs.

What was once a thriving radar school of thousands of men shrunk to accommodate fewer than two hundred by the end of March 1946. Instructors consolidated radar classes and eliminated all but post-war courses: radar mechanic Army Air Force, radar repairman airborne equipment and radar mechanic Ground Control Approach.

Because the future of the Boca Raton Army Air Field was uncertain and its closing a possibility, command headquarters ordered an inventory of all supplies and equipment. But by April 1946, no real progress had been made. The Army Air Force Training Command then sent out the following order: "The main objective of this field is to get the equipment ready for audit." For the next three months, all personnel did just that.

On June 14, 1946, the base underwent inspection by the Department of the Inspector General, Headquarters, Technical Training Command. The recommendation was to close the base. But apparently that wasn't the final word.

New young Florida recruits sent to Georgia for induction into the service failed to arrive at a rate of between 8 percent and 14 percent. On June 3, a conference held

Colonel Rosenham Beam became post commander, replacing Brigadier General Winslow Morse in July 1946. *Photo courtesy of the Boca Raton Historical Society.*

on at the Boca Raton Army Air Field included a discussion on using the facility as a processing and induction station for troops entering the service from South Florida. Plans also included the expansion of the runways in order to keep the field a functioning post-war air base.

NEW COMMANDER NAMED

Brigadier General Morse left in July and Colonel Rosenham Beam became post commander. Under his leadership, reorganization continued and budget cuts became widespread, further reducing civilian support personnel.

By September, the base became an effective recruitment center. Six hundred young men entered the service that month alone. The population of the base went from two thousand in the first quarter of the year to more than six thousand by the end of September.

As recruits arrived and assignments were made, many GIs found themselves in positions for which they had not been trained nor had any interest. However, every now and then, one landed in the right spot. Such was the case with sixteen-year-old Louis Delgado, who arrived in October 1946.

I was inducted into the Army Air Force on July 5, 1946. After indoctrination at Fort Sheridan, Illinois, I went to Lackland Air Base in San Antonio, Texas for basic training. Upon completion of six weeks of training, groups of us would assemble on the Physical Training field for announcements of assignments. A large group of us were put on a train and sent on a two or three-day trip to Boca Raton, Florida for permanent assignment. We arrived in Boca Raton at night in what seemed like monsoon weather in October 1946. After a long train trip of eating mostly cheese and crackers, a heavy downpour and the muddiest conditions one could imagine greeted us. The barracks that we were assigned to did not offer a sense of relief. I remember thinking, "What did I get myself into?" We were assembled in a large gymnasium to receive our assignments. The fellow in line in front of me was told that he was going to be an MP. When my turn came, I said that I didn't want to be an MP, as though I would have had a choice. However I was assigned to the Post Engineers, which turned out to be an assignment that I really enjoyed and would later suit my life's work.

As part of the engineering department, Louis Delgado chauffeurs fellow soldiers in a base Jeep to their next job assignment. *Photo courtesy of Louis Delgado.*

Civilians enjoy tours of aircraft and buildings at the Boca Raton Army Air Field during an open house. *Photo courtesy of the Boca Raton Historical Society.*

Post engineers maintained the base. This included operating the hospital steam boilers, paving and repairing roads, keeping the canals between the barracks and the collection areas clear of lily pads, operating the garbage dump as well as performing other related duties that kept the base running. Delgado learned to operate vehicles, large and small, and according to Delgado, "for a seventeen-year-old, that was great." On Sundays when everyone else played softball, he learned to operate a roller used to pave roads, a skill that subsequently became his assignment whenever there was road paving or repair work to do. With this assignment, he didn't have to do the shoveling and spreading of the asphalt; however, at the end of the day, the roller had to be taken to maintenance compound. At about three miles per hour, it took a long time to navigate the large base.

"I then had to find a way to get to the Mess Hall, and by the time I got there, everyone else had eaten chow and I had to settle for leftovers, so, I don't know if I really had the best assignment," Delgado recalled. "Whenever our supply of road paving gravel needed to be replenished, a group of us would drive dump trucks to a gravel pit about eight miles away, usually on a Sunday, and we would make continuous round trips all day."

OPEN HOUSE

In the summer of 1946, commanding officers held an open house and invited the general public to tour the base. Aircraft, buildings and even some radar equipment were displayed.

ANOTHER HURRICANE

An October 7 hurricane had a minimal effect on Florida and Boca Raton; however, planes were flown from the base to safer locations as a precautionary measure.

The Category 1 hurricane with winds of eighty miles per hour came ashore near Bradenton, south of Tampa. Some piers, wharves and warehouses were damaged, trees were downed and there were brief power outages, but the storm's biggest effect was to the citrus farmers who lost $5 million in crops.

CLOSING OUT 1946

The Boca Raton Army Air Field saw a lot of changes in 1946. Uncertainty surrounding the future of the base affected all personnel, and the changing and restructuring of duties contributed to the overall uneasiness of operations.

Almost five hundred civilian jobs were cut, leaving a gap in support operations. This resulted in untrained personnel being transferred into these positions and on-the-job-training serving as the order of the day. Whole sections were consolidated and reorganized, and while radar instruction still existed, by the end of the year, only 3,895 students attended classes

The year ended on an upbeat note, however, when soldiers held Christmas parties and rang in the New Year in the officer's club and in the white and Negro service clubs.

1947 Time Capsule

The opening session of Congress is televised for the first time • The first U.S. television soap opera *A Woman to Remember* is broadcast • In Ohio, BF Goodrich manufactures the first tubeless tire • Bell Laboratories develops the transistor, a solid-state electronic component

World News

• India becomes independent of Great Britain and is divided into dominions, Pakistan (Muslim) and India (Hindu).
• The first of the Dead Sea Scrolls is discovered.
• The United States gives up on the China peace effort.
• The French break off talks with Ho Chi Minh in Vietnam.
• The United Nations partitions Palestine into two sections.
• The first Aloha Week Parade is held in Hawaii.
• Britain's Princess Elizabeth marries Duke Philip Mountbatten.

National News

• A merger of the *Times* and the *Sun* creates the *Chicago Sun-Times* in Illinois.
• Congress passes the Taft-Hartley Act over President Harry Truman's veto, severely curtailing the power of organized labor.
• Everglades National Park in Florida is established.
• In Long Island, New York, builders erect Levittown, a middle-class suburb; by 1970, more Americans will live in the suburbs than in cities.
• The gross national product of the United States begins its historic post-war surge, ushering in an era of economic growth.

Sports

• World Series champion—New York Yankees • U.S. Open golf champion—Lew Worsham • Pro football champion—Chicago Cardinals • Indianapolis 500 winner—Mauri Rose • Stanley Cup winner—Toronto Maple Leafs • NCAA basketball champion—Holy Cross • College football champion—Notre Dame • Heisman Trophy winner—John Lujack from Notre Dame

Music

• "The Anniversary Song"—Dinah Shore
• "Ballerina"—Vaughn Monroe
• "Chi-Baba Chi-Baba"—Perry Como
• "Mam'selle"—Frank Sinatra
• "Managua, Nicaragua"—Guy Lombardo
• "Near You"—Francis Craig
• "Open the Door, Richard"—Count Basie
• "Peg O' My Heart"—Three Suns
• "Temptation"—Red Ingle
• "White Christmas"—Bing Crosby

Movies

• *Gentleman's Agreement*—Academy Award winner
• *Dark Passage*
• *Black Narcissus*
• *Miracle on 34th Street*
• *The Bishop's Wife*
• *The Ghost and Mrs. Muir*

Cost of Living

New house	$6,650.00	*Food*	
Average income	$2,854.00 per year	Granulated sugar	$.85 for 10 pounds
New car	$1,290.00	Vitamin D milk	$.78 per gallon
Average rent	$68.00 per month	Ground coffee	$.55 per pound
Tuition at Harvard University	$ 420.00 per year	Eggs	$.23 per dozen
Movie ticket	$.55 each	Fresh baked bread	$.13 per loaf
Gasoline	$.15 per gallon		
U.S. postage stamp	$.03 each		

1947

DECISION MADE TO CLOSE THE BOCA RATON ARMY AIR FIELD

By 1947, the need to recruit and train troops for combat was no longer necessary In large numbers all across the country, military personnel suddenly became civilians. Headquarters assessed the need for bases and slated many for closure.

By the end of the first quarter of 1947, the number of military personnel at the Boca Raton Army Air Field dropped to 2,595, and by May, the dreaded decision came down: deactivate the base. All departments would move to Keesler Field in Biloxi, Mississippi, by November 30, 1947, and preparations commenced for the transfer of equipment, materials and personnel.

MORE TROUBLE IN PARADISE

On May 26, 1947, local resident Denver Brittian received a subpoena from the judge advocate's office. Several years earlier, the government had classified Brittian "exempt" from the Selective Service System because of the need for his "indispensable" maintenance skills at the Old Floresta officer's housing area. Now he was going to serve his country in a different capacity—as a witness in a court-martial.

According to official documents, Gladys Jean Merritt, a waitress at Zim's Restaurant, who resided in Dormitory 5, was getting ready for bed around 2:00 a.m. when she heard a knock on the door. A man said he wanted to talk to her. She recognized the voice as that belonging to an earlier customer, First Lieutenant John H. Duhamel, an officer stationed at the Army Air Corps Technical School of Radar. She told him to go away and he left, but she later saw a man outside the window at Dormitory 4, a fact corroborated by a second witness, Jeanette Mitchell.

Denver Brittian, pictured with his wife, Frances, was called upon to testify in a court-martial on the Boca Raton Army Air Field. *Photo courtesy of Arlene Owens.*

Mrs. Frances Werner, who was visiting her daughter, had been sleeping in Dormitory 4 for quite some time when someone came into her room. The person stood beside her bed and said, in a man's voice, "What's the matter, honey? Turn over, I want to get in bed with you." Mrs. Werner screamed and turned on the light only to see the man run from the room.

Brittian testified that at 3:30 a.m., his barking dog woke him. He got up and looked out the window to see a man by his car in his yard. He went out and asked the man what he was doing. The man told Brittian that he had too much to drink and was trying to find his way back to the main gate, a half mile away.

The next day Brittian followed tracks that led from the dormitory area to his property. Later that day, he happened to see the man in the station hospital. He identified him as the individual he had seen earlier that morning in his yard.

First Lieutenant Duhamel was charged with wrongful and unlawful entry of a girl's dormitory (violation of the Ninety-sixth Article of War), but because it was dark and no one could positively identify him as being the individual who was in the dormitories or at the windows, he was not convicted. He was, however, reprimanded and had to forfeit $100 per month for three months.

The F-84 "Thunderjet" arrives at the Boca Raton Army Air Field. *Photo courtesy of the Boca Raton Historical Society.*

FIRST JET ARRIVES

In 1947, the base saw the arrival of its first jet, the F-84 "Thunderjet." During its service life, the F-84 became the first United States Air Force jet fighter able to carry a tactical atomic weapon. It gained its greatest renown during the Korean conflict by attacking enemy railroads, bridges, supply depots and troop concentrations with bombs, rockets and napalm.

HURRICANE GEORGE, THIS TIME A BIG ONE!

In the autumn of 1947, two hurricanes hit close to Boca Raton. The first—referred to as George by the Weather Bureau forecast office in Miami, which was still working in conjunction with the military—was a Category 4 and swept ashore near Fort Lauderdale on September 17. Winds of 155 miles per hour were observed at Hillsboro Lighthouse for one minute, and five-minute sustained winds clocked in at 121 miles per hour.

George was a large hurricane covering virtually all of Florida south of Brevard County. Winds of 100 miles per hour extended 70 miles from the eye wall. Hurricane-force winds covered an area of 240 miles, and winds of 50 miles per hour extended 300 miles.

The slow-moving storm churned at less than 10 miles per hour and crossed Florida with little loss of strength. Before it exited the state near Naples, it dumped heavy rainfall, up to 10.12 inches at St. Lucie Lock. Winds of 100 miles per hour battered the west coast of Florida from Everglades City to Punta Gorda.

Tides rose eleven feet along the coast from Fort Lauderdale to Palm Beach. Widespread flooding occurred, and tornadoes were spotted throughout the state.

President Harry Truman declared a state of emergency along the entire southeast coast of Florida. Seventeen people died, and damages totaled almost $32 million, with one-third attributed to agricultural losses. The hurricane later regained strength in the Gulf of Mexico and came ashore a second time in New Orleans, Louisiana, as a Category 2.

Louis Delgado was at the Boca Raton Army Air Field when Hurricane George hit:

Prior to the hurricane which hit Boca Raton in September of 1947, we would get storm warnings and we would have to board up the windows and doors on our barracks. When the alert was over, we would take down the boards, and then we repeated the same procedure when the next alert was issued.

A soldier stands in thigh-high water in front of a radar-training building. The standing water was left by Hurricane George. *Photo courtesy of the Boca Raton Historical Society.*

The post stockade collapsed under the intense storm. *Photo courtesy of the Boca Raton Historical Society.*

When the alert came for the hurricane, I was at the naval air station in Fort Lauderdale. An announcement was issued that all women expecting childbirth were to report to the base hospital in Boca Raton. Since all of the officers were on duty in Boca Raton, I was ordered to take one of the expectant mothers from Fort Lauderdale to Boca Raton in their personal car. When we arrived in Boca Raton, the storm was imminent and I had no way to get back Fort Lauderdale, so I had to stay in Boca Raton. Once again I survived on cheese and crackers, while back at Fort Lauderdale they had all the good food they could want. After the hurricane, we were busy with restoring the base back to as normal as it could get.

Jim Vaughter was also on the base when George hit: "I was in one of the L-shaped buildings, which were classified as a permanent-type structure. Standing at a window, I watched a wall of a wood warehouse rip off board by board, leaving the two-by-four studs standing.

"My cot was located in the corner of the L by the furnace. I had just gotten up to go to the latrine when the brick chimney came crashing down on the cot and flattened it. They say the winds were 155 [miles per hour] but I thought they were 180!"

Roofs flew off, buildings ripped apart and many of the barracks were lifted from their foundations and flattened. Damages in excess of $3 million hastened the closing of the base.

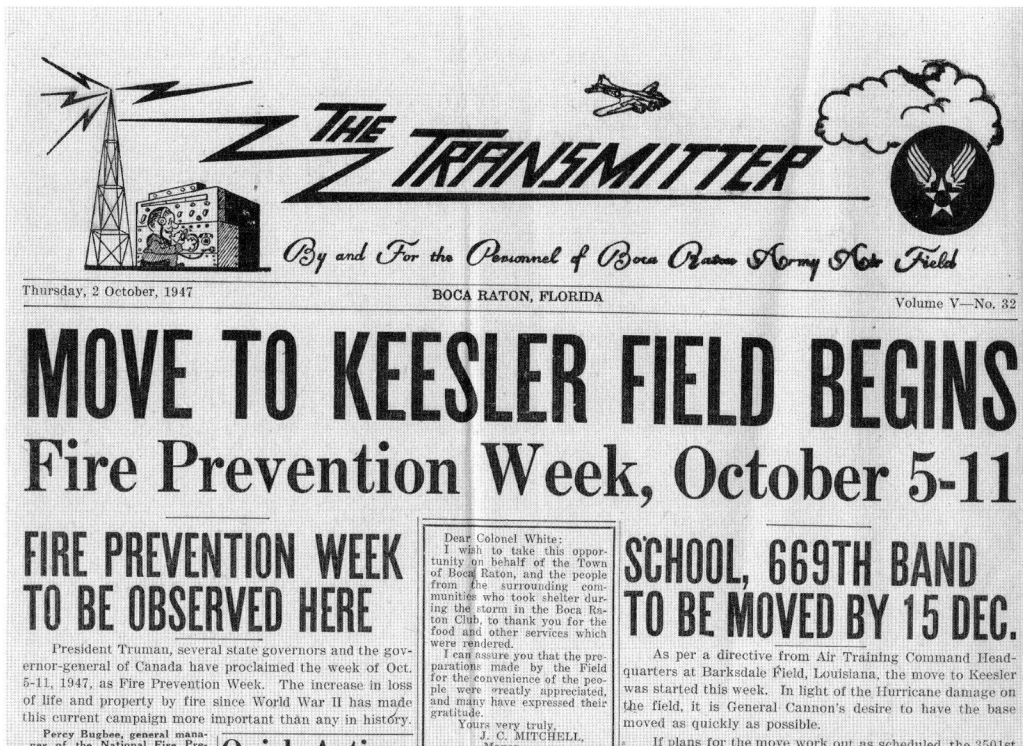

The October 2, 1947 *Transmitter* announces the move to Keesler Field. *Photo courtesy of the Boca Raton Historical Society.*

ANOTHER HURRICANE—KING

To add insult to injury, on October 12, a second hurricane, King, struck Cape Sable on Florida's southwest coast and exited at Pompano Beach—once again grazing the Boca Raton Army Air Field with hurricane-force winds.

King started as a tropical depression off the coast of Nicaragua and meandered north gaining little strength. But after crossing Cuba, it rapidly intensified to a Category 4 storm with winds of up to 150 miles per hour just before it made landfall.

After moving across the state, the storm hit Boca Raton with winds of 95 miles per hour. Flooding caused the worst damage. Six inches of rain fell in seventy-five minutes in Hialeah, Florida, leaving standing water more than six-feet deep. In Boca Raton, Old Floresta residents, whose homes were used to house officers, waded in eight inches of water.

THE BASE CLOSES

For five years, the Boca Raton Army Air Field carried out the secret mission for which it had been designed. Now, however, it was time for its gates to close. Service personnel shipped out to other locations, and only a small contingent of officers and enlisted men stayed behind to lock the doors.

What had once been a proud training facility planted on the cutting edge of military technology had now become obsolete. Jim Vaughter recalls the troops' last hurrah:

> *Myself plus some ten other officers and several airmen that were left at the base inventoried the base and closed up the books for the audit.*
>
> *Knowing that the balance in the officer's club funds had to be sent to Washington to go into the General Fund, we decided to have a big bash and dance at the hotel. Flaming food was served, and we had a waiter for each couple. [Actor] Danny Kaye was at the hotel that night.*
>
> *I remember leaving the hotel around 2:00 a.m. on a Sunday morning when a young man drove into the garage with a car that had a hood a mile long. To the attendant I heard him say, "I want the car washed and a tune-up. Have it ready by 8:00 a.m.!"*

CLOSING OUT 1947

The Boca Raton Army Air Field opened amid fanfare in 1942 yet closed with a whimper in 1947, its clandestine assignment complete. During that time, close to one hundred thousand troops transitioned on and off the field, but those voices would now give way to silence. Yet as quiet returned to the town of Boca Raton, another covert military project began. One that was so secret, it wouldn't be discovered and reported on until 2001.

BOCA RATON
AFTER THE WAR

The abandoned airfield left a multitude of structures intact that became the catalyst for growth in Boca Raton. While the airport returned to civilian hands, land and buildings that once housed thousands of troops were put up for sale for a fraction of the original construction cost. Many of these structures housed thriving businesses while others housed hundreds of workers and their families. This then, is a recap of what happened to the remaining Boca Raton Army Air Field structures and the land this proud airfield once occupied.

AIRPORT RETURNS TO BOCA RATON

In 1948, Mayor J.C. Mitchell notified the War Assets Administration, the Civil Aeronautics Administration and the U.S. Engineers that the town wanted the base and the Boca Raton field "in its entirety" for a municipal airport. Mitchell also heard from a number of businesses owners who inquired about leasing former base buildings.

On December 29, 1948, under the Surplus Property Act of 1944, a quitclaim deed transferred control of the 1,250-acre site containing the base airfield from the United States of America to the Town of Boca Raton. However the deed came with a restriction—the property could be used only as a public airfield until such time as the Civil Aeronautics Administration deemed it was no longer needed for that purpose.

The August 5, 1948 edition of the *Pelican,* Boca Raton's earliest newspaper, stated: "It is believed that the acquisition of this deed by Boca Raton will be one of the greatest steps forward in its entire history toward the building of our town."

A quitclaim deed, dated April 24, 1959, transferred the property from the Town of Boca Raton to the State Board of Education of Florida, thus for a time, this body owned and operated the Boca Raton airport. Also, around this time, the Civil Aeronautics

Administration agreed to earmark 1,000 acres for educational purposes, leaving 201 acres for the public airport.

In a deed dated November 10, 1970, airport property was transferred from the State Board of Education of Florida to the Board of Trustees of the Internal Improvement Trust Fund of the State of Florida. And in a subsequent deed dated January 22, 1973, the property was transferred from the State of Florida Board of Trustees of the Internal Improvement Trust Fund to the State of Florida Department of Transportation.

The Boca Raton Municipal Airport Web site states: "Control of the airport then passed through five different government agencies before the Airport authority was developed to take control of the airport.

"On October 27, 1983 (Lease Agreement No. 3265) the Land Lease became effective from the State of Florida Board of Trustees of the Internal Improvement Trust Fund to the Boca Raton Airport Authority until midnight January 22, 2073."

Managed by Ken A. Day and a staff of five, the Boca Raton airport is publicly owned and designated as a general aviation-transport facility. A seven-member authority, appointed by the Town of Boca Raton and Palm Beach County Commission, governs the facility.

TOWN PURCHASES LAND

In a resolution passed on June 27, 1949, Boca Raton agreed to pay the army $251,184 for 2,404 acres of the former base. Excess land not needed directly by the town would be sold. To pay for the property, the town sold 1,079 acres of this land east of the El Rio canal but retained the water system and its easements; the sewer collection and outfall systems; and many of the winding streets.

Property north of Palmetto Park Road was sold to Joe L. Moore who opened a real estate office, Boca Raton Hills, in February 1950. He offered for sale sixteen warehouse buildings along the Florida East Coast railroad tracks for as little as $1 per square foot and homesites for $5 per front foot on paved roads. He also offered for sale acreage and several other large buildings.

The Reverend Ira Lee Eshleman was interested in purchasing land and buildings for a winter Bible conference. Having turned down a gift of 160 acres at the western edge of the airfield for being "too far out in the boondocks," Eshleman came into town and met with J.C. Mitchell and Joe L. Moore.

"Moore told me he'd been offered $50,000 from a gambling syndicate from Detroit for five acres and one of the buildings, but he didn't feel right about it. I offered him the same $50,000, but said I wanted two buildings and thirty acres. He asked me what my terms would be and I told him, 'As little down as possible and the rest over the longest period I could get,'" said Eshleman.

Moore laid out the terms: $5,000 down, $20,000 in six months and he would finance the balance. Eshleman agreed, and the deal was reached. After much debate over the name, the complex was christened Bibletown USA.

A Ground-Floor Pre-Development Opportunity
To Buy
War Surplus, Masonry Buildings
WITH AMPLE LAND
IN
BOCA RATON HILLS
BOCA RATON, FLA.
A $25,000,000.00 DEVELOPMENT

City water
City sewer
Paved streets
Electricity
High Ground —
the highest elevation and the best
drained property in the country.

Keep Your Eye On Boca Raton Hills!
Watch Boca Raton Grow!

JOE L. MOORE & ASSOCIATES
Developers
Office: Approx. 3 Blks. West of Downtown Boca Raton, Fla.
PHONE 5401

Joe L. Moore opened the Boca Raton Hills real estate office in 1950 and sold warehouses and homesites. *Photo courtesy of the Boca Raton Historical Society.*

On the land Eshleman bought was the officer's club and the men's service club. The former became Bibletown's headquarters and the latter was converted into a sixty-room hotel. Later he bought another forty acres that included the base post office and the theater. The post office became a twenty-five-room lodge for conference staff, and the base theater became the Harry A. Ironside Memorial Auditorium. It housed Eshleman's radio studio when he bought WBDF in Delray Beach. The large base administrative building became Moody Hall, named after the famed evangelist D.L. Moody. It was used as a cafeteria and apartment house until the 1980s.

Eshleman received twenty more acres as a gift from Moore and developed them into fifty lots. He later sold them and gave the proceeds to Bibletown. When Eshleman left Bibletown USA to become chaplain of the National Football League in 1967, the name of the complex was changed to Boca Raton Community Church.

While none of the original base structures remains on the church campus, several building foundations were used to construct the current buildings.

Of the other Boca Raton Army Air Field buildings that were sold by Moore, J. Deutch purchased the headquarters building at the main gate and converted it into a sixteen-unit apartment building. It is now called the Boca Heights apartments.

The two school buildings—T-603 and T-604, one-half mile north on Main Highway—were purchased by Jean Maschuch and Vance & Baune, and converted into apartments.

Theater Building 2, located at Second Street, was sold to J.T. Harvey and operated as a theater.

Purchasing one of the base buildings, the Bo-Del Printing Company moved its plant from Delray Beach to Boca Raton.

Scranton Metal Casket Works acquired one warehouse building, and F.N. Oldstore Inc., a millworking plant from Fort Lauderdale, purchased another large building.

The base administration building became Moody Hall, a cafeteria and apartment house on the Bibletown USA campus. *Photo courtesy of the Boca Raton Historical Society.*

Sydney Rabonovich bought the refrigeration plant on the FEC tracks and restored and conditioned it for the Colonial Packing Co., a meat packing plant.

Other companies that bought warehouse property from Moore included the Wats Manufacturing Company, which made tire changers for aircraft and heavy trucks, and the Sun and Sea Paint and Varnish Company.

OTHER BUYERS

Dominic C. Jalbert established an aerology laboratory in an old base building. His inventive airfoils changed the concept of kite and parachute design.

The Sjostrom Machine Company established operations in an old H building. The company manufactured textile machinery.

The First Church of Christ Scientist purchased one of the base chapels and relocated it to Delray Beach. Because of its size, the building was cut in half before being moved then reassembled on site. An article in the *Delray Beach News* dated January 21, 1949 stated: "The chapel, erected by the government in 1942 at a cost of $30,000, was sold to the local Christian Science Church for $1,500 and moved here by Leonard Bros. Storage and Transportation Company of Miami."

THE MYSTERIOUS "FARMERS" AND SECRET BIOLOGICAL TESTING

Although the Boca Raton Army Air Field became deactivated in 1947, the military didn't pull out completely. Bob Sloan, a longtime resident of Boca Raton, recalled an encounter with several mysterious "farmers" in the 1950s.

Sloan was hanging out at the A&W with a friend when several men walked in. He recalled: "They told me they were farmers but they looked like they were right out of a New York street gang."

Later, in 2004, when Sloan learned that the base was used as a secret biological research center, he remembered the encounter with the "farmers." Through a female friend who dated one of the "farmers," he was able to locate "Duane" (not his real name). Sloan phoned Duane that night.

"He was shocked that someone would be calling him from Boca Raton after fifty years," said Sloan.

Duane told Sloan that he had never even told his wife of forty-eight years what he did in the service and he wasn't going to talk about it now. However, he did admit he was stationed in Boca from 1951 to 1954 and that he was one of eight "farmers" along with a unit of about fifty men. He also said whenever he traveled, a Federal Bureau of Investigation agent was with him, trying to get Duane to slip up and say something about the secret testing. But Duane assured Sloan, "I never said anything to them or anyone else."

Duane said his training was at Fort Meade, Maryland, a place Sloan remembered from his military days as being filled with "highly secretive and classified operations." Sloan also recalled, "I know that they had top scientists, microbiologists and biochemists working on experiments."

Dominic Jalbert flies one of his kites. *Photo courtesy of the Boca Raton Historical Society.*

The Sjostrom Machine Company purchased one of the concrete-block buildings formerly used for radar training. *Photo courtesy of the Boca Raton Historical Society.*

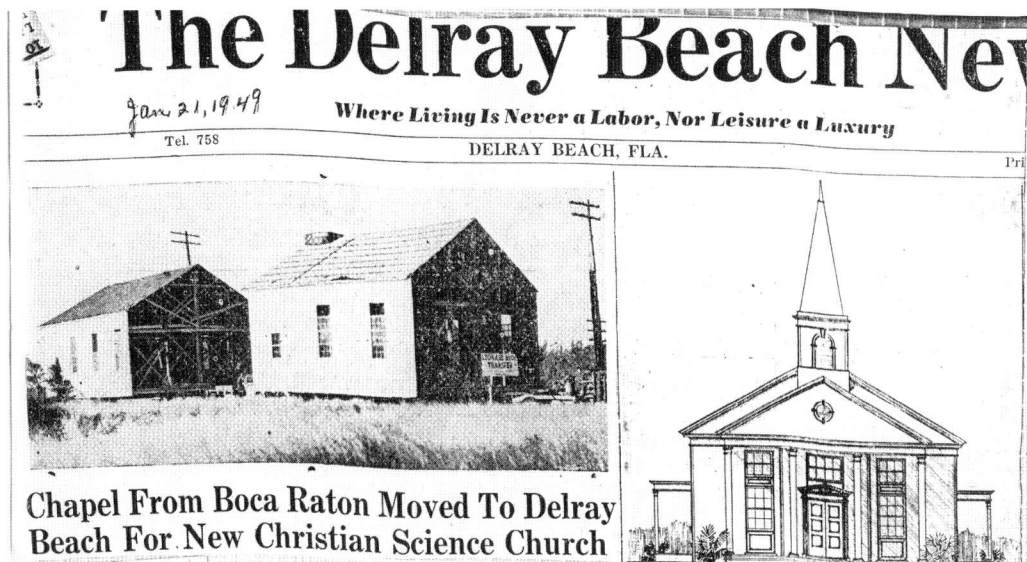

This January 21, 1949 article in the *Delray Beach News* shows how the church was cut in half and moved from the Boca Raton Army Air Field to its present location in Delray Beach. *Photo courtesy of the First Church of Christ Scientist Delray Beach.*

SERVICES
WOMAN'S CLUB BUILDING
Sunday: 11:00 A. M.
Sunday School: 9:30 A. M.
Mid-week services are being held tempo-
rarily on Thursday night at eight o'clock.
Reading Room: Bon Air Hotel building.
Week days 10 to 4 and 7 to 9, except Thurs-
day evenings.

FIRST CHURCH
OF CHRIST, SCIENTIST
DELRAY BEACH, FLORIDA

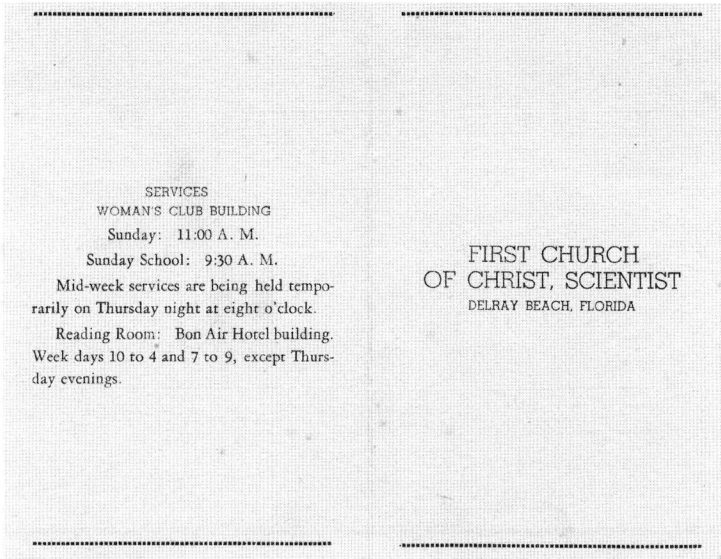

This 1948 church bulletin, from the First Church of Christ Scientist in Delray Beach, explains to the congregation the terms of sale of the base chapel they purchased from the Boca Raton Army Air Field. *Photos courtesy of the First Church of Christ Scientist Delray Beach.*

Report On Building

The present plans for securing a building for this church are to purchase an Army chapel at Boca Raton Field, move it to Delray Beach and renovate it for a permanent church edifice. If we are successful in getting the chapel, it can be made ready for use and dedication probably within a year.

The investment required to complete the building as we would like it, will be about $30,-000, including the lot, cost of moving, landscaping, interior furnishings, organ and some exterior and interior alterations. The completed building will seat 300 persons without enlargement and will represent a value of about $50,000. The Army chapel is constructed of excellent material. These chapels are being sold for church purposes, only, at about five per cent of their original cost or $1,500. An amount of money somewhat less than half the full amount of $30,000 will be required within the next few months but the remainder of the work can be done later.

A sum of $3,300 in cash and bonds is now available in the building fund and another portion may be derived from the sale of the business building formerly used as a church. Application will also be made to the Cornish Will. New city zoning regulations prevent use of the lot owned by the church on N. E. Seventh Avenue and First Court for a church building. Therefore, the membership has voted to sell the lot and acquire another site.

Contributions towards the realization of our building program will be gratefully received. If the Army chapel cannot be secured plans will be made for erection of a building as soon as practical.

BOARD OF DIRECTORS.

Feb. 1, 1948

The First Church of Christ Scientist is located at 200 SE Seventh Avenue in Delray Beach, Florida. A portico and new windows were added to the front. *Photo by author.*

In a July 7, 2002 article titled "Veteran Recalls Secret Weapon Work at Boca Airfield," Eliot Kleinberg of the *Palm Beach Post* wrote about the Boca Raton Army Air Field's use as a secret, biological-weapons testing site. The following is a summary of that story.

After the base closed, a small contingent of military personnel remained behind to produce a secret biological weapon—stem rust of rye, a variation of wheat rust that "can destroy more wheat and other grass crops in less time than any other crop malady." Kleinberg, who did extensive research on the topic, wrote that the wheat fungus could be used on Russian wheat fields to "render wheat stalks barren and starve civilians."

Operating from a tightly controlled installation occupying eighty-five acres just north of the Boca Raton Army Air Field from 1952 to 1957, scientists grew the wheat on plots between and along runways at the old Boca Raton airfield in an area the size of one or two football fields. Fungus spores were sprayed on the wheat, then in a few days, the multiplied spores were harvested and stored in two-gallon, vacuum-sealed stainless-steel containers.

"The idea was to dust the spores onto chicken feathers, then load them with a bomb of sorts that would activate about 500 feet off the ground," Kleinberg wrote.

The spores would adhere to the wheat and allow it to grow, but not produce any grain. The hope was to force Soviet leaders to concede in the event of a war or move resources to feed their starving civilians. The program was shut down in 1957.

Sergeant Tom West, who worked on the project and who was interviewed extensively for the article, stated, "I was in the Army and my idea was to fight the enemy and the enemy at the time was Russia."

Kleinberg and Senator Bill Nelson, a Democrat from Florida, pursued this topic through the U.S. Army Corps of Engineers, who, back in 1994, assembled a report on the activities at the base. In March 2003, Nelson was given a briefing and reviewed appendix K of this report but is unable to reveal its contents, which are considered "classified" information. Senator Nelson was satisfied, however, that his main concern—that no harmful materials had been left behind—had been addressed.

Duane's specific involvement in this project still remains a mystery.

BOMBS AWAY

In the early 1950s, Arlene Brittian-Owens, remembers her yard was home to a leftover souvenir from the Boca Raton Army Air Field. After being told it was safe, the family used the old bomb as a planter.

Visiting his granddaughter, Arlene Brittian-Owens, Anthony Fagen inspects a leftover souvenir from the Boca Raton Army Air Field. *Photo courtesy of Arlene Owens.*

The Boca Raton campus of Florida Atlantic University opens with fewer than eight hundred students in 1964. *Photo courtesy of Florida Atlantic University.*

FLORIDA ATLANTIC UNIVERSITY (FAU)

In March 1960, an agreement between the Civil Aeronautics Administration and the state went into effect, requiring the state to establish a university by 1969 on one thousand acres of the former Boca Raton Army Air Field.

Spearheading a campaign to raise $100,000 for planning and architectural fees, Thomas F. Fleming Jr., who had led Boca Raton's campaign to bring the university to the base site, enlisted the aid of Farris Bryant, Democratic gubernatorial candidate, who pledged to find construction funds.

Bryant went on to become governor of Florida and led the issuance of $25 million in bonds to build the Boca Raton campus and aid other state institutions to upgrade their physical plants.

Vin Mannix, in a December 16, 1979 *Boca Raton News* article "etcetera by mannix and the news" wrote: "The ground was broken for FAU in December 1962, and what was once a bastion of winged military might became an institution of high learning in 1964."

The campus officially opened on September 14, 1964 with fewer than eight hundred students. Today, five campuses—Boca Raton, Davie, Downtown Fort Lauderdale, Jupiter and Port St. Lucie—serve an enrollment of twenty-five thousand graduate and undergraduate students. Eight colleges offer seventy-six undergraduate majors along with sixty-eight masters programs and doctoral degrees.

BASE WAREHOUSES SURVIVE OFF CAMPUS

Next to the Florida East Coast railroad tracks, several of the original base warehouses currently accommodate a variety of businesses. The same Dade County pine post-and-beam work used in several of the original base buildings can be seen in each of the structures. Originally constructed as individual warehouses, post-war additions connect the buildings.

Douglas Wheeler, president of Wheeler's Moving & Storage Inc., manages eight of the buildings east of the spur. He said the buildings were well constructed,

1940 *Photo courtesy of Remote Sensing Lab, Dr. Charles Roberts, associate professor of geography, Department of Geoscience at Florida Atlantic University, USDA CJF-14-53.*

1943 *Photo courtesy of Remote Sensing Lab, Dr. Charles Roberts, associate professor of geography, Department of Geoscience at Florida Atlantic University. Army Air Corps from the National Archives, record group 373, can 6A-730, no. 9.*

1964 *Photo courtesy of the City of Boca Raton.*

1995 *Photo courtesy of Remote Sensing Lab, Dr. Charles Roberts, associate professor of geography, Department of Geoscience at Florida Atlantic University. USGS NAPP 6966 121.*

This aerial view is of the Florida Atlantic University campus was taken in 2004. *Photo courtesy of FAU Office of the University Architect and Vice-President.*

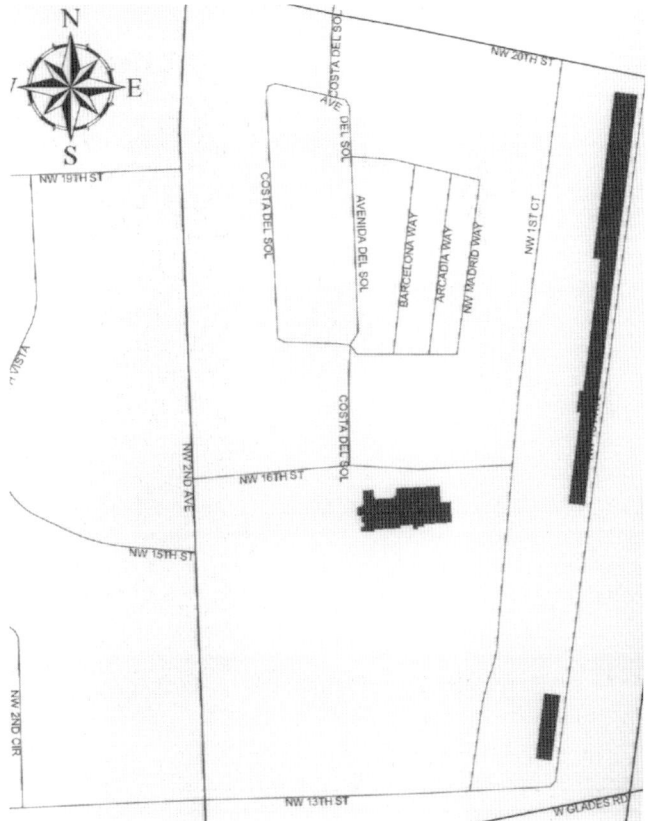

This map indicates current businesses located in warehouses and a former radar-training building near the Florida East Coast railroad tracks. *Photo courtesy of Dick Randall, City of Boca Raton.*

This page and opposite: A variety of businesses now occupy warehouses that were once part of the Boca Raton Army Air Field. *Photos by author.*

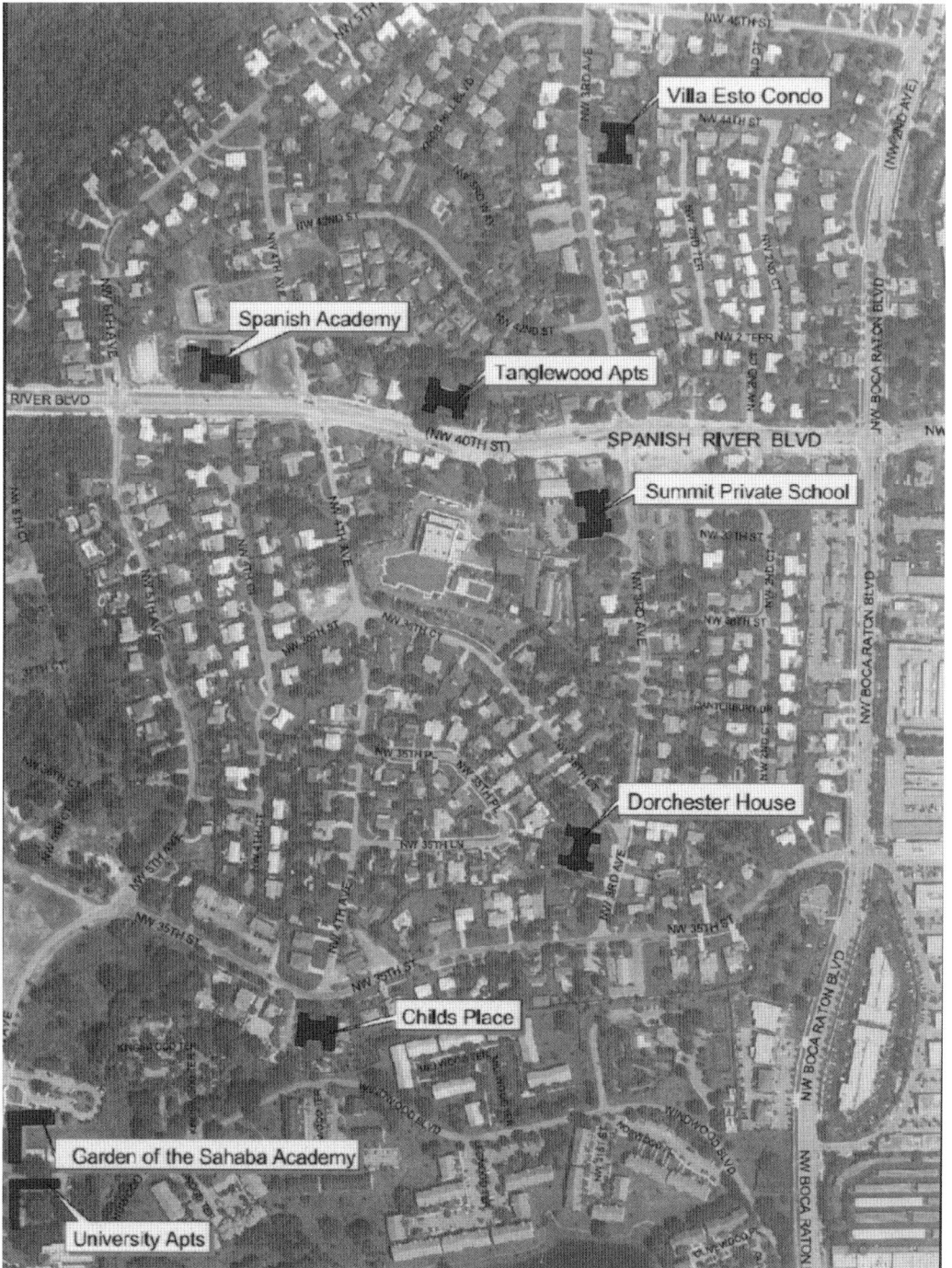

Ten former H-shaped radar-training buildings and two L-shaped officer's quarters were renovated after the base closed and are currently being used as schools, apartments and businesses. *Photo courtesy of Dick Randall, City of Boca Raton.*

noting that warehouses built more recently tend to have an inside temperature of at least ten degrees higher in the summer than the temperatures inside the base warehouse structures.

TEN BASE BUILDINGS SURVIVE OFF CAMPUS

Off campus, many of the H and L buildings were converted to apartments, schools and businesses. The following are the names and addresses of the structures existing as of April 2005:

SCHOOLS

Left: Spanish Academy, located at 4475 Spanish River Boulevard. *Photo courtesy of Diane Bradford.*

Right: Garden of the Sahaba Academy, located at 3100 NW Fifth Avenue. *Photo courtesy of Diane Bradford.*

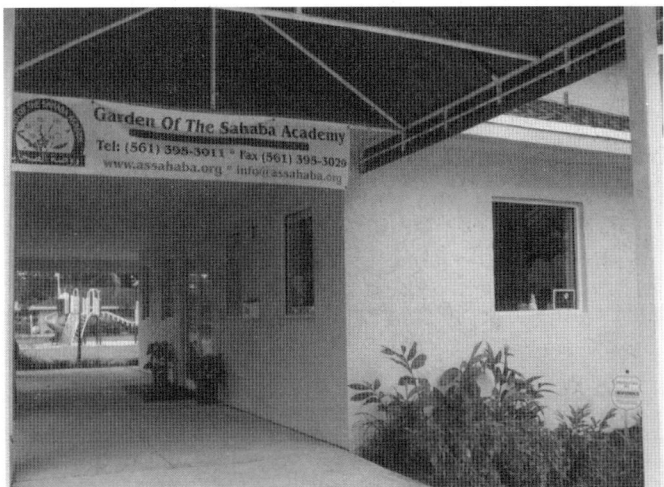

Not pictured: Summit Private School, 3881 NW Third Avenue; A Child's Place, 414 NW Thirty-fifth Street.

APARTMENTS

Boca Heights Apartments, located at 101 Pine Circle [Headquarters Building]. *Photo courtesy of Diane Bradford.*

Tanglewood Apartments, located at 325 Spanish River Boulevard. *Photo courtesy of Diane Bradford.*

Villa Esto condominiums, located at 4300 NW Third Avenue. *Photo by author.*

Left: Mizner Oaks Apartments, located at 398 NW Twenty-second Street. *Photo courtesy of Diane Bradford.*

Not pictured: University Apartments, 3000 NW Fifth Avenue; Dorchester House, 3535 NW Third Avenue.

THE SECRET VAULT

In the spring of 2005, Mizner Oaks Apartments was in the process of being sold. Plans were to raze the structure to make room for sixty contemporary apartments. Before that happened, however, the apartment's owner, Cortlandt Schuyler, revealed that the structure had a vault in the interior of the building.

Upon inspection, the vault measured ten feet by twelve feet, had poured concrete walls and a seven-foot-tall steel door. The door itself was gone but the steel frame was still intact and a plate on the upper mantle read "York Safe and Lock Co. York, PA." Without the door, it was impossible for experts to tell much about the safe, however, Seymour Sendrow, a radar supply sergeant, offered insight into its use: "In it we kept classified records and schematics, parts of secret equipment and information on nonexpendable supplies such as sensitive radar equipment."

NINE BASE BUILDINGS SURVIVE ON CAMPUS

Nine original Boca Raton Army Air Field structures remain on the Florida Atlantic University campus. One of the historically significant buildings, known as T-3, sits on a portion of seventeen acres of Phase II of the FAU Research Park, property leased to HRM Limited for development. At 19,200 square feet, T-3 is the largest building and was the base engineering shop. Using post-and-beam construction, it was built from Dade County pine, a wood no longer available.

Currently, T-3 is being retrofitted to accommodate future businesses. Plans are to maintain the architectural integrity of the post-and-beam interior and include a small museum dedicated to the Boca Raton Army Air Field.

A sister building, T-4 that sat just south of T-3, was demolished in late 2004. At 9,800 square feet, it was used for base administration.

T-30, a relatively small building surrounded by foliage, sits on the north apron of the old runway. During the war, it was used to store flammables. Hazardous materials from

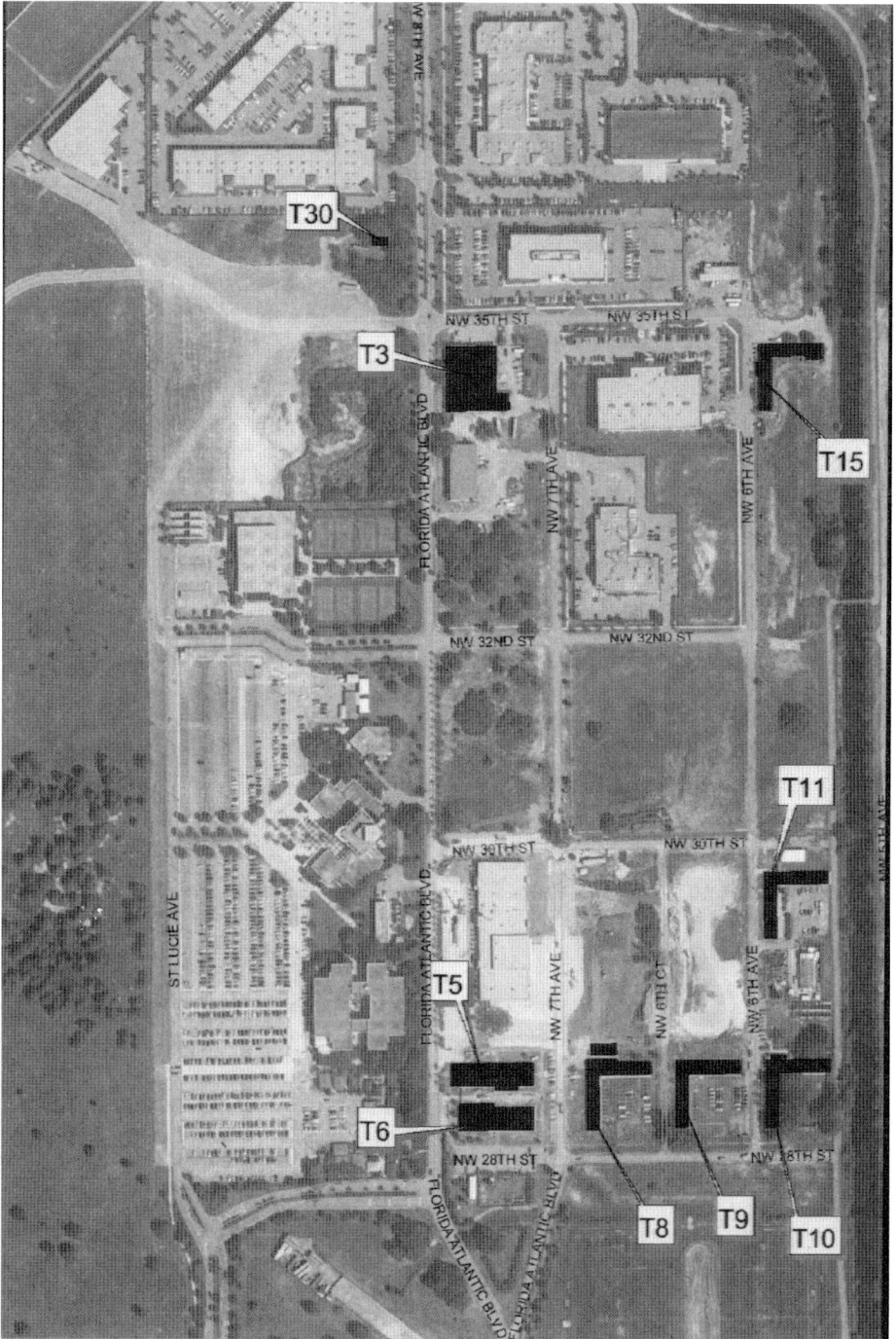

Nine former Boca Raton Army Air Field buildings still stand on the campus of Florida Atlantic University. This map shows their locations as of March 2005. *Photo courtesy of Dick Randall, City of Boca Raton.*

At 19,200 square feet, the T-3 building is the largest of the remaining buildings and housed radar equipment. It sits on the FAU Research Park. *Photo courtesy of Diane Bradford.*

Businesses will be housed in the T-3 building, but plans are in place to maintain its integrity by exposing part of the interior posts and beams. *Photo by author.*

The T-30 building stored flammables on the Boca Raton Army Air Field. Under lock and key, it now secures the university's hazardous-waste materials. *Photo courtesy of Diane Bradford.*

The T-6 building was once used as officer's quarters. It currently houses labs and workshops. *Photo courtesy of Diane Bradford.*

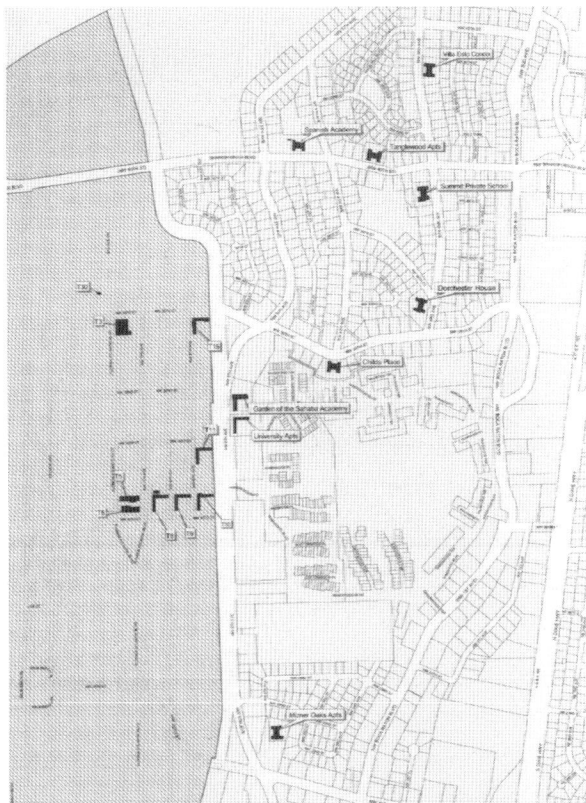

This map shows the remaining Boca Raton Army Air Field structures (except for warehouse area and Boca Heights Apartments) that are being used by Florida Atlantic University and area businesses. *Photo courtesy of Dick Randall, City of Boca Raton.*

Florida Atlantic University's labs are currently stored in this building under lock and key. A chain link fence surrounds the structure with access through a locked gate.

The other seven remaining buildings on the campus were being used for classrooms, labs and offices. From 1942 to 1947, the rectangular-shaped T-5 and T-6 buildings were used as school buildings and the L-shaped buildings—T-8, T-9, T-10, T-11 and T-15— were officer's quarters.

In September 2004, Hurricanes Frances and Jeanne struck a double blow to several of the buildings, damaging their exteriors and rendering them unusable.

THE SANBORN WALL

In early 2005, Dr. Peter Barrett visited Boca Raton and discovered that the wall between the former Villas, where he had lived, and the Sanborn home, where the Germans came ashore, was still standing. Though hidden from the road by tropical foliage, the poured concrete wall is located along a beach access path on the south side of the Beresford Condominium on A1A, south of the Palmetto Park Pavilion. It is distinguished by its tiered construction, rectangular decorative spaces and large round finial. Several photos of the original wall built adjacent to the Sanborn home in 1937 are in the Boca Raton Historical Society collection.

Left: Sue Gillis, archivist for the Boca Raton Historical Society, holds a 1944 aerial photo of the Boca Raton Army Air Field. She is part of a group working to preserve the remaining wartime buildings on the Florida Atlantic University campus. *Photo by author.*

Right: Bonnie Dearborn is the administrator of the South Florida Region Preservation office, Division of Historical Resources, Florida Department of State. She stands in front of what was once the T-4 building. The 9,800-square-foot building was demolished in late 2004. *Photo courtesy of Diane Bradford.*

Left: Arthur Jaffee, World War II veteran and Ario Hayams Endowed Chair Professor at Florida Atlantic University's Southeast Wimberly Library, co-authored a master plan with Mike Singer for the preservation and use of the remaining World War II buildings. *Photo courtesy of Diane Bradford.*

Right: Florida Atlantic University student Amy An, former president of the Xi Omega chapter of the Phi Alpha Theta national history honor society, has rallied students to support the preservation of the Boca Raton Army Air Field buildings. *Photo courtesy of Diane Bradford.*

HONORING HISTORY:
THE BOCA RATON
HISTORICAL SOCIETY

The Boca Raton Historical Society—along with a groundswell of World War II veterans, Florida Atlantic University students, preservationists, concerned groups and individuals—is working to preserve the remaining World War II buildings on the FAU campus. Known as the BRAAF Heritage Society, it is their desire to see the past and future coexist in harmony by convincing university planners to design additional classroom space in conjunction with the original wartime structures. The group also hopes to create a park and museum dedicated to the Boca Raton Army Air Field and the veterans who served there.

To find out more about this project and how you can help, visit: www.bocaratonarmyairfield.com.

BRAAF VETERANS

BELL, WILBUR. *Cadet, 1943.*
Bell became an engineer; he initially worked for Westinghouse then formed his own company. At eighty-five, he is retired and currently resides in Maine.

BENDER, DR. CHARLES. *Radar student, 1945–1946.*
After being discharged in May 1946, Bender attended medical school. He lives and practices medicine in Florida.

BUTLER-CHAVEZ, BERNICE. *Nurse, 1943.*
Butler married Manuel Chavez, an officer stationed at the BRAAF. She and Manny have been married sixty-three years and live in Florida.

CARSWELL, ARCHIE. *Transportation, 1942–1943.*
Carswell served as a private professional chauffer until he retired in 1987. He currently resides with his wife, Irene, in Florida.

CHAVEZ, LT. COL. MANUEL "MANNY." USAF (RET.) *Pilot, 1943–1945.*
Chavez married Bernice Butler, an Army Air Corps nurse stationed at the BRAAF. He later went into the intelligence field and the military diplomatic service in Latin America. He retired in 1966 as a Lieutenant Colonel and now lives in Florida.

COCHRANE, JOHN H. *Radar student, 1945.*
Cochrane went back to school on the GI Bill and eventually worked in the old Bell System. He lives in South Florida where, he says, "I can sit on my balcony and look southwest at the signal beacon at Boca airport, home to me so many years ago."

CORNWELL, MARION "PINT." *Radar student, 1944.*
A former civil engineer, Cornwell lives with his wife, Nell, in Florida. He is eighty-one.

DAVEY, ROBERT. *Radar student, 1944.*
Davey managed several brokerage firms. When he moved to Florida he opened Davey Realty, Inc., Davey Management, Inc., and Davey Property Services, Inc. At seventy-six, he resides in Florida with his wife, Marie.

DELGADO, LOUIS. *Post engineer, 1946.*
Delgado attended Purdue University and received a BS degree in Engineering Technology. Later he received an MBA from Loyola University. He is currently retired and lives in Illinois.

EDDINGER, WILLIAM. *Radar student and instructor, 1942–1946.*
Eddinger moved his wife, Phoebe, and daughter, Patty, to Boca Raton where he served as superintendent of public utilities. He died in 2001.

GANSER, HUGH. *Radar student and instructor, 1943–1944.*
After the war, Ganser went into the family business, Ganser Home Improvement Company, a roofing and siding company. He retired in 1981 and lives with his wife, Georgiann, in Wisconsin. He is eighty-seven.

GERBER, RICHARD. *Radar student and instructor.*
A graduate of the University of Florida, Gerber went into the mortgage and real estate business. He is still active in his profession in Virginia.

GODETT, NELSON. *Flight officer, bombardier, 1945.*
Godett became a pediatric pedorthist in Pennsylvania. He retired to Florida with his wife, Ruth, in 1987.

JONES, GERALD. *Cadet, 1943.*
Jones returned to the University of Nebraska to further his studies as a mechanical engineer. He currently lives in Florida.

KANTOR, PHILIP. *Radar student, 1945.*
Kantor went into wholesale and retail furniture sales and retired in 2002. He is seventy-nine and currently resides with his wife, Elyne, in Florida.

MAGAFAN, HARRY. *Radar student, 1942–1943.*
Magafan became a food distributor until his retirement in 1984. At age eighty-six, he lives with his wife, Irene, in Florida.

MALAVARCA, ED. *Radar student, 1944.*
After the war, Malavarca went into the packaging industry. He moved to Florida in 2002 and at eighty-six is semi-retired.

MANGRUM, JOHN. *Cadet, 1943.*
Mangrum was an Episcopal priest for fifty-five years. He retired in 1990 and at age eighty-three, currently lives in Florida.

MAGUIRE, ANN. *Nurse, 1943.*
After the war, Maguire served as a private nurse to millionaires. She bought a small bungalow in Boca Raton where she retired. She is deceased.

METRO, PAUL. *Radar student, 1944–1945.*
Metro retired in 1986 from the Continental Insurance Company, where he worked as a fire and workers' compensation safety consultant. Now eighty-two, he lives in New Jersey.

MIKUNDA, LOUIS. *Cadet, 1943.*
Whereabouts unknown.

MIZE-JACKSON, EDITH. *Flight Nurse, 1943.*
After the war, Mize-Jackson moved to Florida and became a freelance writer. Whereabouts unknown.

ORNSTEIN, JOE. *Radar student, 1943.*
Ornstein became a salesman for RCA. He retired after thirty years with the company and currently resides in Florida with his wife, Ida Rosen.

PHEARS, WILLIAM. *Cadet, 1943.*
Phears earned a PhD in engineering and became deputy commissioner of public works in Hempstead, New York. He retired to Florida and remained there until his death in 2003.

RAY, WAINO. *Radar instructor, 1942–1947.*
Ray taught high school English then joined Travelers Insurance until his retirement in 1977. He currently lives in Maine with his wife, Doris. He is eighty-eight.

RENNER, NED. *Radar student, 1946.*
Renner served as company clerk with the 9th Air Force in Erlangen Air Base in Germany. There, he had the historic opportunity of going to Nuremberg where he witnessed the trial of the Nazi war criminals. Renner made the U.S. Army a career as an army criminal investigation division agent and polygraph examiner. He retired in 1966 as the agent in charge of the Pittsburgh, Pennsylvania, field office, U.S. Army

Criminal Investigation Division. At seventy-six, he currently lives in Maryland with his wife, Gloria.

ROBISON, BILL. *Radar student, 1945.*
Robison worked as a mechanical engineer in the defense industry. He later went from the private sector to the military sector at the Picatinny Arsenal, Dover, New Jersey. He is semi-retired and lives in Florida with his wife, Helen.

ROMERO, CHARLIE. *Guard, 1943–1944.*
Retiring from the roofing business at age seventy-six, Romero currently lives in New Mexico. On July 3, 2004, he was inducted into the Amateur Athletic World Hall of Fame Museum in a ceremony held at the Albuquerque Convention Center.

SENDROW, SEYMOUR. *Radar supply sergeant, 1944–1945.*
Sendrow became a lab technician with Dupont then opened Sendrow Men's Apparel in New Jersey. Now retired, he and his wife, Pearl, live in Arizona. He is eighty-five.

SHULMAN, RALPH. *Cadet, 1943.*
Shulman went back to work as executive vice-president for Transocean, an importer of area rugs. He retired in 1982 and currently lives with his wife, Zelda, in Florida.

THIELE, ALVIN. *Cadet, 1943.*
Colonel Thiele went on to join the 416th Bomb Group in the 9th Air Force in Europe. Whereabouts unknown.

VAUGHTER, JIM. *Sales officer, 1947.*
Whereabouts unknown.

WALLISH, ED. *Radar student, 1946.*
After the war, Wallish became a computer programmer. He retired in 1994 and currently resides in Illinois with his wife, Christine. He is seventy-five.

WEISS, GERALD. *Radar student, 1944.*
Weiss became a Chrysler automobile dealer. He retired to Florida in 1980 where he lives with his wife, Alice.

WICKERT, PETER. *Crew chief, 1945.*
Whereabouts unknown.

WILLIAMS, JAMES. *Cadet, 1943.*
Williams became a member of the original Tuskegee Airmen, the first black aviator group to

serve in the U.S. Armed Forces. He also became one of 101 black officers at Freeman Air Field placed under arrest in 1945 for refusing to sign a regulation that would keep the officers' clubs on the base segregated. Though he was not jailed, a letter of reprimand went into his file and remained there until 1995 when it was removed by the air force. His refusal, and that of the other black officers, proved instrumental in bringing the issue of segregation in the military to the forefront.

Williams attended Creighton Medical School in Chicago, earning an MS degree in medical surgery in 1956. He established a medical clinic on Chicago's south side with his brothers, Jasper and Charles, and in 1971 became chief of surgery at Chicago's St. Bernard Hospital. When civil rights leader Martin Luther King Jr. was in Chicago, Williams served as his physician.

Williams was named chief of surgery at Doctors Hospital of Hyde Park (the old Illinois Central Hospital) in 1992 and in 1996 the hospital's surgical suite was named in his honor. Now retired, Williams, eighty-five, lives with his wife, Willeen, in New Mexico. They have been married fifty-four years.

OTHERS

BARRETT, DR. PETER V. *Spotter.*
Barrett lived in Boca Raton until 1946, when his family returned to California. He attended undergraduate school at UCLA and Harvard Medical School. Since 1967, he has been in academic medicine at UCLA. He is currently working part time and is responsible for geriatric medicine at Harbor-UCLA Medical Center. He is seventy.

DAVEY, COL. KENNETH W. USAF (RET.) *Pilot.*
Davey became a USAF career officer. His last assignment was as commander of the 6921st Security Wing, stationed in Misawa, Japan. His second career was as advisor with the College of Business at the University of South Florida. Now retired, he lives with his wife, Ana, in Florida. He is ninety.

DURHAM, ELDRA. *Mail clerk, 1943–1947.*
Durham served the post office for fourteen years and was a volunteer at Broward Medical Center in Fort Lauderdale for twenty years. In 1988, she retired from service to her community. She lives in Florida where she has been a resident for seventy years; she is eighty-six. Her husband, Richard, who also served as a mail clerk at BRAAF, is deceased.

KOBAYASHI, TOM. *Son of Japanese farmer displaced by BRAAF, 1942.*
Kobayashi was drafted in 1945 and spent several months in Okinawa as a medic with the occupation troops who guarded Japanese prisoners of war. In 1952, he graduated from the University of Miami with a liberal arts degree. He continued with the family landscape business until he retired in 1997. Now seventy-eight, he lives in Florida with his wife, Maizel.

BIBLIOGRAPHY

PRIMARY SOURCES

ARCHIVES
Boca Raton Historical Society
File headings:
   "Boca Raton Army Air Field"
   "Women in the War Years"
   "Air Force Base"
   Scrapbook of Lt. Col. Frank F. Fisher
   Scrapbook of S/Sgt. Stanley A Biedron

Minutes. Special Meeting of the Town Council of the Town of Boca Raton, Florida. November 5, 1941.

Order awarding Possession, In the United Sates District Court in and for the Southern District of Florida, Miami Division. May 16, 1942.

Staff Judge Advocate's Review. General Court Martial Orders. No. 210. June 5, 1947.

Subpoena for Civilian Witness. Boca Raton Army Air Field. May 26, 1947.

INTERVIEWS AND PERSONAL COMMUNICATION

ORAL AND WRITTEN MEMORIES

Barrett, Dr. Peter
Bell, Wilbur
Bender, Dr. Charles
Butler-Chavez, Bernice
Carswell, Archie
Chavez, Lieutenant Colonel Manny, USAF (Ret.)
Cochrane, John H.

Cornwell, Marion
Davey, Colonel Kenneth W., USAF (Ret.)
Davey, Robert
Delgado, Louis
Durham, Eldra
Ganser, Hugh
Gerber, Richard
Gillis-Hanson, Gladys
Godett, Nelson C.
Jakubek, Pat
Jones, Gerald
Kantor, William
Kobayashi, Tom
Magafan, Harry
Malavarca, Ed
Mangrum, Rev. John
Metro, Paul
Mikunda, Louis

Ornstein, Joe
Ray, Doris
Renner, Ned
Richardson, Porter
Robison, Bill
Romero, Charlie
Rolle, Henry Van
Sendrow, Seymour
Shulman, Ralph
Simmons, Nora
Sloan, Bob
Thiele, Alvin
Vaughter, Jim
Wallish, Ed
Wells, Arthur A.
Weiss, Gerald
Wickert, Peter
Williams, Dr. James B.

E-mail Correspondence

Boomhower, Ray. Indiana Historical Society. E-mail message to author, June 9, 2004.

Hagedorn, Dan. Smithsonian National Air and Space Museum, Archives Division. E-mail message to author, June 24, 2004.

Houghton, Walter E. Acting Deputy Director, Fort Lauderdale-Hollywood International Airport. RE: Ground Control Approach, 2005.

University of Miami Athletic Office. E-mail to author June 19, 2004.

Secondary Sources

Books

Army Air Forces Training Command Year Book. Boca Raton, Florida, 1947. (Available at the Boca Raton Historical Society, BRHS.)

Barnes, Jay. *Florida's Hurricane History*. Chapel Hill: University of North Carolina Press, 1998.

Buderi, Robert. *The Invention That Changed the World*. New York: Simon & Schuster, 1996.

Conant, Jennett. *Tuxedo Park: A Wall Street Tycoon and the Secret Palace of Science that Changed the Course of World War II*. New York: Simon & Schuster, 2002.

Craven, Wesley Frank and James Lea Cate. *The Army Air Forces in World War II*. Vol. 6, *Men and Planes*. Chicago: University of Chicago Press, 1949.

Curl, Donald W. and John P. Johnson. *Boca Raton: A Pictorial History*. Virginia Beach, Virginia: The Donning Company, 1990.

———. *Palm Beach County*. Northridge, California: Windsor Publications, 1986.

Johnson, Stanley. *Once Upon A Time: The Story of Boca Raton*. Miami, Florida: Arvida Corporation, 1979.

Kleinberg, Eliot. *War in Paradise: Stories of World War II in Florida*. Cocoa, Florida: The Florida Historical Society Press, 1999.

Phears, Bill. *Ain't But It Can Be*. Melboourne Beach, Florida: Blue Note Publications, Inc., 1993.

Thuma, Cynthia. *Boca Raton*. Images of America. Charleston, South Carolina: Arcadia Publishing, 2003.

DOCUMENTS

Eshleman, Dr. Ira Lee. *A Gold Coast Miracle, "Great Things He Hath Done."* Florida, privately printed.

Grub Frieser, Diana. *Boca Raton Airport Title Documents*. City of Boca Raton, January 10, 2003.

Hand, L.C. "Report on Securing Chapel." First Church of Christ, Scientist, Delray Beach, Florida, January 6,1948.

Lloyd, Joanne M. *Yankees of the Orient: Yamato and Japanese Immigration to America*. Thesis, Florida Atlantic University, 1990. (BRHS)

Patterson, Dorothy W. *Yamato/New Town Project*. Delray Beach, Florida: Expanding & Preserving Our Culture Heritage, Inc., 2003.

"Report on Building." Delray Beach, Florida: First Church of Christ, Scientist, February 1, 1948.

Rowe, Francis A. *Background and Position Paper*. January 10, 1980.

NEWSPAPERS AND PERIODICALS

Arjemi, Craig. "World War II flight nurse featured in special issue of Life magazine." *Florida Nursing News*, July 7, 1985. (Available at the Boca Raton Historical Society, BRHS.)

"Army Air Force 'Invaded' Boca Raton in 1942." Unknown source. August 13, 1953. (BRHS)

"Army Man Recalls Days of Air Base Here." *Boca Raton News*, June 1, 1965. (BRHS)

"Boca Field To Get 'Chute Landing School." *Delray Beach Times*, June 20, 1944. (BRHS)

Boca Raton Historical Society, 2003 Annual Gala. *Sentimental Journey*, 2003. (BRHS)

Brown, Drollene P. "World War II in Boca Raton: The Home Front." *The Spanish River Papers* 14.1 (Fall 1985). (BRHS)

"Chapel From Boca Raton Moved To Delray Beach For New Christian Science Church." *The Delray Beach News*, January 21, 1949.

"Colonel was a 'big spender.'" *Boca Raton News*. May 26, 1974. (BRHS)

Davis, Rick. "The Fight for Equality." *Creighton University Magazine*, Summer 2000.

"A Special Study of Operation 'Vittles.'" *Aviation Operations Magazine*, April 1949.

Holst, Ella Elizabeth. "The Life of a Boca Raton Woman." *Spanish River Papers*, Spring 1986. (BRHS)

Kelly, Dick. "Joe College to Take Place of G.I. Joe." *Fort Lauderdale News*, June 19, 1961. (BRHS)

Kleinberg Eliot. "Veteran Recalls Secret Weapon Work at Boca Airfield." *Palm Beach Post*, July 7, 2002.

———. "Military Refuses to Detail Cold War Testing in Boca Raton." *Palm Beach Post*, June 21, 2003. (BRHS)

Lynfield, Geoffrey. "Yamato and Morikami: The Story of the Japanese Colony and Some of Its Settlers." *The Spanish River Papers* 8, no. 3 (Spring 1985). (BRHS)

"Nine Fliers Meet Death In Crash." *Delray Beach Times*, May 12, 1944. (BRHS)

Remember When...A Nostalgic Look Back In Time. Seek Publishing: 1940, 1941, 1942, 1943, 1944, 1945, 1946, 1947.

Strasser, Cindy. "1940's—The War Years." *The Spanish River Papers* 17 (August 1993). (BRHS)

Waldeck, Jackie Ashton. *Boca Raton: A Romance of the Past*. Boca Raton: The Bicentennial Committee of Boca Raton, 1981. (BRHS)

———. "How Boca Helped Win the War." *Boca Raton Magazine*, March/April 1982. (BRHS)

"Whatever happened to...Boca Raton Army Airfield." *Boca Raton News* December 16, 1979. (BRHS)

"Women in Boca Raton Fifty Years of History." *The Spanish River Papers* 17 (August 1993). (BRHS)

Wylie, Philip and Laurence Schwab. "The Battle of Florida." *Saturday Evening Post*, March 11, 1944.

MICROFILM

Boca Raton Radar School. Roll Numbers: B2053, B2054, B2055, B2056, B2057, 2058, 1942–1947. (BRHS)

INDEX

ABOUT THE AUTHOR

Sally J. Ling is a freelance writer and marketing consultant (www.sallyjling.com). She is a special correspondent for the *Sun Sentinel* newspaper and her work has appeared in *Gold Coast*, *Delray Beach*, *Boca Life* and *Carolina Bride* magazines.

She is former owner and CEO of MedSearch Network, a medical recruiting company, and has served as marketing director for a number of corporations. When not writing, she helps businesses develop and implement strategic business and marketing plans.

She lives in Deerfield Beach, Florida, with her husband, Chuck.